The Red-tailed Hawk

THE GREAT UNKNOWN

Map by Jonathan White.

Jonathan White is a historic cartographer, portolan artist and illustrator specialised in creating beautiful naval maps such as the one created specifically for this book. Your source for the highest quality hand-made maps and nautical illustrations, movie set/props, publishing, marketing, gaming, retail sales and collecting.

To enquire about commissions and obtain additional information please visit Jonathan's website at www.indianbay.net

The Red-tailed Hawk

THE GREAT UNKNOWN

Beatriz E. Candil García

© Copyright 2008 Beatriz E. Candil García.

All Rights reserved. No part of this publication may be reproduced, stored in a retrieval system, or transmitted, in any form or by any means, electronic, mechanical, photocopying, recording, or otherwise, without the written prior permission of the authors.

The right of Beatriz E. Candil García to be identified as the author of the Work has been asserted to in accordance with the Copyright, Designs and Patents Act 1988.

First edition printed in Spanish language Victoria, Canada in 2004.
Second Edition published by Yarak Publishing, 2007.

This English Language Edition was edited and translated into English from Spanish by Beatriz E. Candil García and Robert Forstag.

A CIP catalogue record for this book is available from the British Library.
ISBN 978-0-9555607-4-3

Printed in England

Book layout and design by Yarak Publishing and Moyhill Graphic Design
Original cover designed by Tomás Brugarolas Villa and adapted by Moyhill Design.

The papers used in this book were produced in an environmental friendly way from sustainable forests.

Yarak Publishing
Nest 585, Office 6, Slington House, Rankine Road
Basingstoke RG24 8PH
United Kingdom

Please contact us for more information regarding our publications, future titles and for submissions of manuscripts.

I would like to dedicate this book finally to my dear family;

To Arjen, my husband, best friend, business partner and dad-to-be of our first baby.

I would also like to dedicate this book to my eagerly awaited baby son (who will surely be a great footballer) and who will be born shortly after this book is published.

I love you both very much!

Red-tailed Hawk

Like a dark blur the hawk drops in a steep dive
Into a small bundle of prey, commands the airwaves.

With its great shoulders, its broad and lengthy wingspan;
Folded in stillness, then, it perches on a branch or pole.

What pools of opportunity here on the open prairie,
This island spinning with food to its electric eyes.

On the grass floor, among the rooted slender stalks,
even in dust few are safe from its sharp talon, its beak of doom.

Look, there, watch it bathe in blue air, soaring and circling,
Small blotch on the mirror of pond, its rusty tail its signature.

Poem by Twyla Hansen
from "PRAIRIE SUITE: A CELEBRATION, 2006"
Spring Creek Prairie Audubon Center

www.springcreekprairie.org

CONTENTS

Foreword by Paul A. Johnsgard 1
Author's Preface 3
1. 1. The Buteo jamaicensis 7
2. Subspecies of the Buteo jamaicensis 55
3. The World of the Red-tailed Hawk 107
4. Equipment 155
5. Basic Training 181
6. Hunting with the Red-tailed Hawk 205
7. Health and Well-being 229
8. Red-tails in Love 241
9. Useful Addresses 251
10. Glossary 255
11. Appendix 269
12. Bibliography 277
    About the author 280
    The Yarak portal 281
    Acknowledgements 282

Foreword

Paul A. Johnsgard

It is probably true that the first hawk that most North American bird lovers encounter after becoming interested in birds is the red-tailed hawk. It is easily the most widespread of the North American buteo hawks, and over much of the country is the species most often seen perched on telephone posts, in large trees, or soaring high overhead. I first encountered it when I was five or six years old in the late 1930s, while visiting my grandfather's eastern North Dakota farm. He pointed out a large flying hawk in the distance and described it as a "chicken hawk." Like other farmers of the time, he wrongly accused it of carrying off poultry, an activity much more likely to be caused in that region by red foxes or great horned owls. In any case, I learned to watch for red-tailed hawks along every country road we travelled by car, and soon realized that these beautiful hawks were not always identical in appearance. During fall migration in late September and October, and again during early spring, many North Dakota buteo hawks otherwise having the shape, size and behaviour of the typical almost white-bellied red-tails were dark brown to almost coal black on their breasts and bellies, and their very dark tails sometimes lacked any of the rusty upper tints that I had come to rely on for red-tail identification. I eventually learned, after I finally had acquired a field guide as a teenager, that these dark seasonal migrant visitors were "Harlan's hawks," then considered a separate species, but now regarded only as a red-tail subspecies. And, some hawks that I saw in North Dakota during the summer were much paler on their upper parts than the usual red-tails, and I also came to know them eventually as Krider's hawks, considered to be a distinctive Great Plains subspecies. With such variations, watching red-tails became much more interesting, say, than observing American kestrels, the other fairly common roadside hawk of North Dakota's Red River Valley.

We are now provided with a book-length description of the red-tailed hawk by Beatriz E. Candil Garcia. It is a revised translation of an earlier Spanish-language book, *El Gran Desconocido*, for which I was happy to provide a few illustrations. It is an interesting combination of information on basic biology and the use of red-tails in falconry, something that most of the raptor references in my library totally ignore, as did I also did in my *hawks, Eagles and Falcons of North America*. I have met only a few falconers, but in my experience they are among the most observant and committed hawk watchers that I have ever encountered.

I think the photographs in this book are especially useful, as they provide an excellent indication of the plumage variations to be found among red-tails, and many interesting aspects of falconry, banding and captive husbandry. Beatriz Candil's bilingual abilities and extensive travels have also put her in touch with hawk-lovers on both sides of the Atlantic, and have given this book a much wider audience and greater value than would be the case for most writers on raptors. Her own personal affection for keeping and handling red-tails, especially her "Bandit," also shines through clearly. I think that this work deserves reading by all those who find the presence of raptors circling gracefully in the sky overhead, or in direct pursuit of wild prey, to be one of the greatest thrills that can be experienced in all of nature.

Preface

Even though the Buteo jamaicensis is one of the most common and abundant birds of prey used in falconry in the United States, it remains relatively unknown in Europe, except for the United Kingdom, where it has been flown for many years.

The red-tailed hawk is one of the best birds that we can use in falconry. As we will see later on, it is possible to fly it as a broad-wing but also as a long-wing or falcon: thus, it is capable of capturing a greater variety of quarry and of employing different hunting techniques. Among the many animals that it can be used to hunt are rabbits, hares, squirrels, ducks, pheasants and even partridges.

It is a bird that can be used not only by beginners, but even by expert falconers. The red-tailed hawk is relatively easy to train, easy to obtain (here in Europe there are a good number of red-tail breeders and in the US it is widespread), and both noble and obedient. It can also be maintained in optimum physical condition without a great deal of difficulty and comes with heavy-duty plumage involving little or practically no imping.

I fell in love with this species many years ago, and I eventually owned, and hunted with, several different red-tails. I have amassed an extensive library on this buteo, as it was my desire to know everything possible about this wonderful bird of prey. The genesis of this volume lies in the reading of the books I've accumulated, as well as in the numerous exchanges I've had with expert falconers and ornithologists. The red-tailed hawk has always fascinated me, in much the same way that they have come to captivate the residents of New York City - , as you will see later. They have a set of unique qualities that, in my view, set them apart from all other hawks.

I originally published this book in Spanish in 2004. It was my intention then, as it is now, to provide falconers and bird of prey lovers with the most complete and up-to-date book possible on this species.

After an initial first step into what began as a hobby in my early years, I found myself involved professionally in the world of falconry and birds of prey, from the first phases of training through to breeding and falconry courses both for adults and children. It was then, in 2004 that I decided to publish what I had learnt as a tribute to a much loved yet quite unknown species in the world of European falconry: The red-tailed hawk. This book is a revised edition of the original Spanish version with additional content and has been the fruit of all my years of experience.

I have great memories of the times I shared with Bandit, my male red-tail and his parents, just as I have of all the other moments I was lucky enough to be able to share with the rest of the birds of prey I then lived with: peregrine, lanner and saker falcons, American and European kestrels, Harris hawks, goshawks, Cooper's hawks and a variety of species of owls.

An unfortunate accident that I suffered forced me to concentrate on writing for a time, something that I have always loved, as well as projects that I had in the back of my head. I decided to travel for a while and moved abroad to the Netherlands where I began my research into an idea for a possible future book, a multilingual falconry dictionary, and shortly after, I moved back to London (UK) where I had grown up as a child. It was here that, in collaboration with my now husband, Arjen Hartman, that we published the dictionary (titled "Ars Accipitraria: An Essential Multilingual Dictionary for the Practice of Falconry and Hawking"). Arjen was also quite passionate about nature and showed a great interest in falconry. Another project also began to take shape: Yarak Publishing was born and so was Yarakweb, the world's only multilingual falconry portal, all with one clear concept in mind which was to enable communication among all who share a passion for falconry and birds of prey.

This book has been written from the heart, and is the product of my most cherished dreams and noblest intentions. Above all else, it is a tribute to the red-tailed hawk. I genuinely hope that this book will inspire in my readers the same kind of passion for this extraordinary bird that has captured my own heart.

Once again, I would like to thank all who have contributed to this book for their love and support.

May you find many red-tails soaring above you – or at your command to guide you in your journey through this life.

Happy red-tailing!

Beatriz E. Candil García.
London, 2008

*Close-up of "Bandit", the author's male red-tailed hawk, seen here with juvenile plumage.
Photo by Arturo Gil-Marzán*

The Red-tailed Hawk

"Ladyhawk", photograph by Charlie Kaiser.

1. The Buteo jamaicensis

Native to the Americas, but still unknown in many areas of Europe, the Red-Tailed Hawk, classified for the first time by the German naturalist Gmelin in Jamaica in 1788, is one of the most common birds of prey—and one of the species most frequently utilised by falconers—throughout North America.

The red-tailed hawk (*Buteo jamaicensis*) is a member of the order of falconiformes, which includes all of the diurnal birds of prey, such as eagles, kites, and falcons, and thus shares many physical characteristics with these related species. It belongs to the genus Buteo, which are rather heavy and stocky birds of prey: somewhat smaller than eagles and with wings that are generally both wide and long. There are 27 species of *Buteo* throughout the world, mainly in Africa, America and Europe, with the common buzzard being one of the most common members of this group.

The red-tailed hawk is a bird of prey that has been categorised within the United States as one of the "broad-winged hawks", which signifies birds of prey having wide wings, and which also includes the common buzzard (*Buteo buteo*), Ferruginous hawk (*Buteo regalis*), Harris Hawk (*Parabuteo unicintus*) and, finally, the red-tailed hawk (*Buteo jamaicensis*). However, certain subspecies of red-tailed hawks really seem to more closely resembles an accipiter than a typical buzzard or buteo.

This species is geographically distributed throughout the American continent (see Chapter 2), from Alaska and Canada, throughout the continental United States, and south into Mexico, extending into the Caribbean, Cuba, and Panama. It is not found outside of these areas except for specimens that have been imported and that are being raised in captivity. The variations of the subspecies of the red-tailed hawk, in addition the different natural hybrids that arise as a result of the cross-breading that occurs between the red-tail and other buteos, are considerable. Thus, the pure specimens of the red-tailed hawk are sometimes very difficult to classify or recognize because of the variation in size and plumage. The juvenile and adult plumage is very similar among all of the subspecies, although adults tend to have wider wings and shorter tails. Otherwise, the only important variations between juveniles and adults are the colour of the tail (which is not yet red), and the chest, which is usually more prominent in younger birds than in adults of the same subspecies.

There are three phases of plumage during the life of the bird: 1.) "down", which is present at hatching; 2.) the growth following the first moult and prior to the growth of complete adult plumage (normally occurs at the age of two years, and is often called "intermewed"); and 3.) adult plumage. In the intermediate stage, the plumage is very similar to what it later becomes at the

The Red-tailed Hawk

adult stage, although the irises are pale and they retain juvenile flight feathers and rectrices.

The male and female chicks have similar plumage, and though generally they differ in size (females usually being larger than males), differences in plumage and size are not always helpful in determining the sex of the bird, as we will see later in this chapter.

The red-tailed hawk gets its name from the colour of the top part of its tail, a hue that may range from orange-red or even faded pint to bright red, and that usually appears one year following birth, or during the second autumn of the bird's lifetime, at the time of the first moult. In some instances, it is not fully visible until the second moult. In the vast majority of adult birds, there is a broad black band (subterminal band) at the end of the tail. The lower part of the tail is usually bright white, although in some instances it takes on a silvery colour.

Generally, these hawks are large in size, although there is wide variation among individual specimens with respect to subspecies and geographical distribution (see Chapter 2). *Still, it is generally true that red-tailed hawks from the West are darker and smaller, and look more like the common buzzards (Buteo buteo) that are seen in Spain and the rest of Europe*, with the tiercels weighing 700–800 grams (24.7–28.2 oz) and the females weighing 1,000–1,300 grams (35.3–45.8 oz), although there are of course specimens that lie outside these ranges. *Conversely, red-tailed hawks from the East are generally larger* with 900g-1.2kg for males (31.7–42.3 oz) and 1.5–2 kg. (52.9–70.5 oz) for females, although there are frequent cases of females that weigh over 2kg (70.5 oz). There are also many males that are as almost as large as females of other red-tail subspecies. *Eastern red-tails are generally paler in colour, and one rarely finds darker specimens in the East (or for that matter, albino specimens in the West)*. As for their relative measurements, the averages for the typical red-tail are as follows: a length of 43–56 cm (16.9–22 in), a wingspan of 109–142 cm (42.9–55.9 in). *Among the red-tailed hawks, it is the fuertesi subspecies that has the widest wingspan.*

The Buteo jamaicensis has large feet that vary according to particular subspecies, and these are quite powerful, being among the strongest possessed by birds used in falconry. Its beak is very strong and extremely sharp, in order to be able to tear the flesh of its prey, and even though unlike falcons, red-tails usually don't bite, they can cause a nasty injury. The point of the beak looks like a needle and the sides of the upper part of the beak are used like knives to cut flesh that is held in the lower part of the beak. In juveniles, the cere is usually a yellow-green colour in youngsters, and it gradually changes to a purely yellowish colour as they grow older.

Its eyes are located in the front part of the head, allowing binocular vision and excellent depth perception. Given that its sense of smell is deficient in comparison to mammals, just like the majority of birds of prey, its other senses

1. The Buteo jamaicensis

Juvenile Buteo jamaicensis. Here we can see the change from the grey phase to a more yellow phase typical of young red-tails. Photograph by Brad. S. Silfies.

In this picture, "Bandit" shows us his nictitating membrane or third eyelid. This translucent membrane can be drawn across the eye for protection when attacking prey but also serves to keep the eyes moist and clean. Nictitating membranes are present in birds, reptiles, amphibians and fish but also in some mammals, like humans although usually no longer functioning. Photo by the author.

The Red-tailed Hawk

Photograph of Canadian falconer Roy Priest at 14 years of age with his first passage red-tailed hawk, his first falconry bird and "master" of this noble art. Roy stills remembers this red-tail with much affection and confessed that they shared unforgettable moments and were inseparable. Photo by courtesy of Roy Priest.

José María Cabrera and his female Buteo jamaicensis calurus, "Beta", one of the stars of the bird of prey display at Safari Madrid.

1. The Buteo jamaicensis

are all the more acute, especially that of sight, which in red-tailed hawks is 8 to 15 times as powerful as that of human beings. Eye colour can serve as an aid in determining the bird's age, as well as its state of health. Yet colour varies among species and may change with age. Juvenile red-tails usually have grey eyes that, over time, change to a yellowish-grey colour before finally becoming dark brown. As is the case with many birds of prey, these eyes, as a means of affording protection against the frequent attacks of the animals they attack, have a third eyelid or nictitating membrane, as can be seen in the upper photograph on page 9.

The Red-Tailed Hawk is one of the birds of prey most commonly used in falconry and displays—both by novices and by recognized expert falconers—outside of Spain, especially in North America (where the species is especially prevalent) and in the United Kingdom.

This hawk has also been used in numerous films, including old Hollywood westerns: whenever the sound used to signify the presence of birds of prey or "eagles" has been used in such productions, it has often been the call of the red-tailed hawk that has been heard. In addition, the bird has been a "co-star" of a number of unforgettable films, such as Ladyhawke, although of course, red-tailed hawks did not yet exist in the thirteenth century, the time during which the film was set.

The red-tailed hawk actually offers many possibilities for use in falconry, as well as in displays. It is relatively easy to train, although it sometimes requires patience—a great deal of patience when it comes to things like maintaining its weight and teaching the bird to make its first jumps from the falconer's fist, something that can become very frustrating in the case of novice falconers working with their first bird. What often happens in such cases is that the red-tailed hawk is very resistant to make "the first big jump" and often thrusts its leg prior to jumping, which can cause some nasty injuries to the falconer. However, it is possible to learn a lot from these birds, who are highly intelligent and who seem at times to even display a certain sense of humour.

Once they have been trained, red-tails can become a true pleasure to work with, and it is possible to establish a very special relationship with them, since they are very friendly and gentle with the falconer, saving their aggression for the quarry of the hunt, at which time they become persistent and times unyielding in their rapid and devious pursuit. They function well with any type of lure, whether it is a dragged lure (rabbit lure) or the swung lure typically used with falcons. If one is patient with them from the very beginning, avoiding making crucial mistakes, the whole ordeal of "making" in after they have caught a prey can also be surprisingly quite simple.

Just as is the case with the majority of hawks, red-tailed hawks usually have a very stable character, providing that they have been parent-reared for later use in falconry and not hand-reared by human beings (i.e., imprinted)—the latter course often entailing dangerous consequences with a such a powerful raptor. This good character is yet another advantage to flying and training this magnificent bird of prey. It is definitely possible to learn great things from these creatures, and to share wonderful times together just as I have.

The red-tailed hawk is not only a bird that can be used for traditional hawking, and will gladly surprise us, allowing us to hunt a wide range of quarry (i.e., small birds, ducks, partridges, pigeons, pheasants, and even feathered geese, in addition to rabbits, hares, squirrels, and even mongooses). This depending of course on the subspecies of red-tail that we have and they prey that are available in our hunting area. In addition, a red-tailed hawk can be used for soar hawking, and therefore is a very versatile hunting bird.

This bird is not currently used a great deal in Spanish falconry, although in recent years this trend has begun to change. The truth is that there are still not too many falconers in Europe willing to try their luck with a red-tail, since there is a certain negative prejudice toward the bird for being a buzzard (i.e., rather than a more traditional falcon or hawk used for falconry) and because of problems that have resulted from the incorrect handling of this bird of prey. "Bandit"—a specimen of Buteo jamaicensis that I once personally owned—was used to train students in falconry courses. Bandit flew both in falconry displays and courses throughout Spain, not only in adult courses but also in courses given to children, and there were never any problems with aggression on its part. A red-tail that has been properly cared for and trained, although displaying marked preference for one or two persons with whom she generally is more comfortable, may fly very well and be safely handled by others without ever displaying aggression —reserving its aggression for pursuing its prey. It is even possible for a genuine friendship to develop between man and bird that transcends its unquestionable usefulness as a hunting partner.

In addition to its excellent hunting abilities, the red-tailed hawk has also been successfully used in falconry flying displays since, like any hawk, it is a thing of beauty to watch in flight, although it must be said that however gentle and well-trained it may be, it is not a Harris Hawk, and is a bird that bears more similarity to an eagle than to a typical hawk or buzzard. Even here in Spain, there are a few displays with the red-tail. One display where they have been used and which I've always enjoyed, is the one that takes place at Madrid Safari, where "Beta" a *B. jamaicensis* female from the western US, has been flown by José María Cabrera and his team. To see a red-tail being flown is absolutely fascinating, and is an experience that one will not soon forget. Whenever I see the red-tail in flight at an event like this, guided by the hand of a trained master, I am seized by a certain melancholy that this bird of breathtaking beauty and grace has not received its due recognition for its unquestionable dominion of the skies.

Diet

The Buteo jamaicensis will easily adapt to available food sources and its diet will vary considerably depending on geographic location, available quarry, weather and season of the year, in addition to other factors. Some red-tails will also establish preferences for certain quarry even if there are other abundant sources of food in the same area. This just goes to show once again how independent red-tails are with regards to individual tastes.

However, taking into account the above, through research carried out, the diet of the Buteo jamaicensis has been reported to be made up of mainly the following:

Mammals: Most of the red-tails's diet is made up of mammals depending on their geographic location, season of the year and availability of different sources. Amongst these we can find cottontails, voles, snow-hares and hares in general, ground squirrels and grey squirrel, rodents in general and also bats. When food is scarce, red-tailed hawks can also feed on carrion and carcasses of cows, coyotes, cattle, deer, etc. Occasionally, if they are very hungry and have no other food sources available they will also eat roadkill. Mammals, however, form the base staple of the diet of the red-tailed hawk and depending on all the above-mentioned factors they can make up an average of 68% of their diet, thus being the food source they consume the most.

Birds: (including quail, pheasant, duck, pigeon, starlings, small birds and even other birds of prey). Red-tailed hawks are one of the largest predators of goshawks in the US, and have been known for also taking domestic fowl or poultry although they have been unfairly nicknamed "chicken hawks" as they sometimes feed on the left-overs from poultry farms. The Cooper's hawk however, does attack poultry. The percentage of birds that makes up the diet of red-tailed hawks in general can vary from 4–58% while the average is approximately around 17.5%.

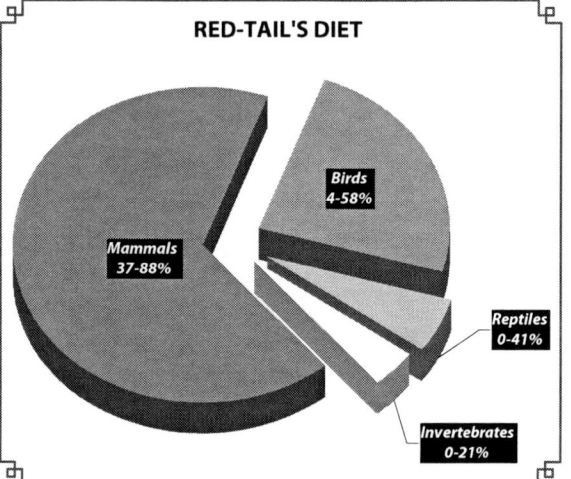

Reptiles: Snakes and lizards make up between 0–41% of their diet, the average consumption being around 7%.

Basically, the above food sources would be the main components of a red-tail's diet in the wild. As we can see, most of its diet is made up of mammals, followed up by birds, reptiles and invertebrates; however, not all red-tails will feed on the same prey. As we have seen, this will depend on their geographic location, availability and abundance of prey (as they can easily adapt to available food sources) and also their natural preferences. **It may seem trivial, but it is important to really be aware of their diet in the wild, as ideally this is what they should be fed in captivity. This is true for any captive-bred bird of prey used in falconry.**

When hunting with our birds or feeding them, we should follow this golden rule and always feed them with prey that they would naturally hunt themselves in the wild. We can also add vitamin supplements to their food, although if we are giving them a healthy and varied diet (see diet above) these will usually not be necessary. These should really only be used occasionally and as an extra help during the moult, breeding season or to help with recovery of any health problems that should arise. Vitamin supplements should never be used on a daily basis as a lack of vitamins is obviously not good for our hawk but an overdose of vitamins can also be detrimental to their health. Also, most importantly vitamins are not absorbed as easily by the body as they are when they are present in natural food sources. That's why getting the right diet is vital for your hawk's health and also for a proper response out on the field.

Red-tails normally enjoy great health and can generally be considered as "tough cookies". This is also one of the reasons that makes them ideal for falconry, especially for beginners, but when training them we should consider a few things; mainly that they do accumulate fat reserves quickly and it can be very difficult (and slow!) to lower their weight if this happens. Having said this, we must keep this in mind and be patient as when we first get our red-tail it may take us a while to get this excess weight down until we can start training and eventually flying our bird. In this sense, another important rule that definitely applies here is to not do things hastily and try to lower their weight as quickly as possible so we can be out in the field sooner. This should never be done with any bird. The saying "good things come to those wait" has never been truer in this sense. We must take a deep breath and be prepared to perhaps spend a longer period of time gradually lowering our hawk's weight (weighing must be done daily!) and carefully watching its diet.

For modern falconers it may not be all that straightforward to get a hold of the usual quarry that red-tails hunt in the wild, but the following easily available food sources, if provided in their correct amounts and varied (never feed your hawk on the same food source alone!) can be useful and can certainly provide a healthy diet:

- Mice and rats
- Rabbit (wild rabbit – caught, not the one we can buy in a supermarket as this is bred for humans and has a lot of fat!)
- Pigeon
- Quail
- Chicken wings
- Frozen day old chicks and turkey poults

The previous diagram, again confirms that mammals are the staple of a red-tail's diet. If we wish to provide a diet which is as similar as possible to what they would eat in the wild and at the same time ensures optimum health, we must weight our hawks daily before feeding them or flying and register this in a logbook (the pages in the Appendix, at the back of the book can be photocopied and used for this). **Remember that any bird's diet, including our red-tail's, must always be as similar as possible to what it would be in the wild.**

Rabbit is also one of the best food sources for our red-tail and one of the main food sources of red-tails in their natural surroundings. Although rabbit can be used as a food to generally help reduce their weight, red-tails do very well with this meat and will usually maintain their weight. We must also vary their diet and include as discussed: chicken wings, which can be crushed in order to provide calcium; day-old chicks which also help maintain their weight; rodents (field mice instead of those we an find in cities to avoid disease) as well as laboratory rats, quail and pigeon (breast or wings, tip: pigeon's wing makes an excellent tiring and should be fed regularly). Falconers may purchase both rats/mice and day-old chicks as well as quail already frozen and supplied in boxes which can be quite cost-effective. Chicken and quail can also be bought in regular supermarkets or butcher's (though they can have more fat on them than those bought from falconry suppliers). Pigeons and rabbit can usually be bought through farms. For a comprehensive list of food suppliers based in your country, please visit the worldwide directory at **www.yarakweb.com**.

Quail is an excellent food source for raptors. If we also crush the bones in a quail leg and feed them with the meat, we will be providing a good intake of calcium. Photo by author.

The Red-tailed Hawk

I have not mentioned squirrel meat because in many countries it is illegal to hunt squirrels but American falconers will confirm that this also is an excellent food source and can provide great slips.

Pork or beef should never be used with any bird of prey or mince and needless to say meat should be as fresh as possible. We should also try to avoid feeding our raptors city pigeons and should this accidentally occur, they should be discarded and thrown directly into the bin, far away from our red-tail and from us. Pigeons carry many diseases which can be passed on to our birds and often are contaminated with poison which has been left to exterminate them. Roadkill or dead animals that we find should never be fed either as we don't know what exactly killed these animals. It is possible that they were ill already and not paying attention to traffic, were run-over.

Anything fed to the red-tailed hawk must first have the head, crop, and digestive system removed first, because these areas represent loci of possible contagious diseases such as trichomoniasis, as well as parasites. It is especially important to remove the head and digestive system also from mammals such as hares and rabbits, for these same reasons as they too can carry dangerous diseases such as myxomatosis.

It is advisable to vary the food sources (see Appendix) and to never continuously provide the same food, even though this is obviously the easier course of action. One-day-old chicks, for example, are an excellent source of Vitamin A and a favourite "easy-to-prepare" food of many

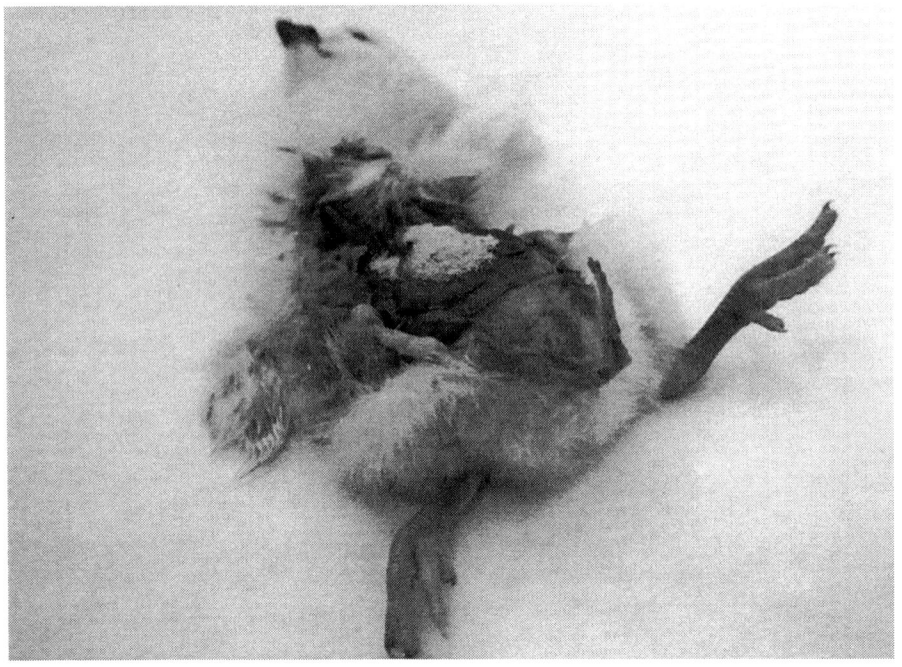

One day-old chick, defrosted, opened up and sprinkled with a vitamin supplement to which water will be added, to help dissolve the powder and for easy ingestion. Photo by author.

falconers, but they have no calcium and the yolks they contain may lead to an excess build-up of cholesterol. Any food provided should be given with the bones crushed (i.e. a chicken wing with the bone crushed inside in order to assure a proper supply of calcium plus feathers or fur and skin (in order to facilitate cleansing of the crop and throwing up of the casting).

It is very important to properly prepare the food in order to prevent the red-tail from choking on bones, and to assure that a quality meal is being provided to the bird. If the food being used has been frozen, it should be naturally defrosted by being removed from the freezer on the previous day, and left to defrost on a tray in the kitchen (or, under very hot conditions placed in the refrigerator in order to prevent the meat from rotting). Microwave ovens or other methods of "quick heating" should never me used to defrost the food. If you want to use food that has been obtained from a hunted animal, first always remove the head and stomach (and crop, in the case of birds) and examine them for parasites. Then place the remainder of the animal in the freezer for at least two hours prior to taking it out and feeding it to your bird. In this way, transmission of trichomoniasis and any harmful parasites will be avoided. Yet another recommendation is, in the case of meat from animals that have not been recently hunted, to warm it with the heat from your own hands, since the red-tail like other birds of prey, enjoys any meat much more when it is served at room temperature, rather than cold from the refrigerator. Remembering this can be very useful, especially in the initial stages of manning and training our bird and we will probably manage to get them to eat sooner with this tip.

Birds of prey don't drink nearly as much as human beings, although they usually do like to drink when they bathe themselves. Therefore, if your bird does not bathe itself frequently like most young birds—or even as a preventive measure— you might want to spray the bird with a plant spray full of water prior to feeding it, in order to prevent dehydration. You will find that some birds will drink this spray off their own feathers. And if this should not happen, there is also the trick of occasionally serving them their food in a bowl of water, although we have to be careful to not let the food stand in the water too long or it will become "washed meat". Additionally you can also just spray the food well in water. You should also make sure that your bird has daily access to a bath so that it can wash itself (and, because it will also drink from this bath, it is important that the water in it be clean). In addition, the bird should have access to a space in the garden where it can sit in the sun: this will facilitate absorption of Vitamin D and help release the natural oils needed to keep their feathers moist— which will in turn facilitate their being able to preen their feathers.

From time to time a vitamin supplement may be added directly to the meat your bird eats or to a little water and then sprayed on the meat. But you should be clear that this is indeed only a supplement to, and not a replacement for, the vital nutrients that the bird can only obtain from being properly fed. There are many brands and different types of supplements out there and the ones I have used are Nekton or Vitahawk, as well as aloe vera but it is well worth also asking your vet.

Many of the animals that are prey of the red-tail are also part of the diet of other birds of prey, such as the ferruginous hawk, the Harris hawk, the great horned owl, all of which share the same habitat as *the Buteo jamaicensis* without fighting over the same prey. For example, the great horned owl and the red-tailed hawk are generally of similar size; it might even be said that they nocturnal and diurnal versions, respectively, of the same predator. But they do not enter into conflict, since they are active during different times of the day. Similarly, the ferruginous hawk is generally active earlier than the Buteo jamaicensis and tends to hunt in open spaces without a lot of trees or "outposts", and therefore has different hunting preferences to the latter.

Although the type of food enjoyed by the red-tail remains the same throughout the bird's life span, the quantity of food that they need does depend on a number of different variables: age, ambient temperature, level of activity, and sex. Accordingly, the bird will typically require more food in the fall and winter than in the spring and summer. At such times, a minimum of 130–150g (4.6–5.2 oz) of food a day is required (the equivalent of one rabbit every 4–6 days). This is the same amount of food as is required by larger birds of prey that are used in falconry.

By contrast, a smaller quantity of food is needed in summer and spring in order for the bird to maintain its body

Photograph courtesy of Nekton (Germany) which illustrates some of the best recommended raptor supplements.

1. The Buteo jamaicensis

Typical image of a red-tailed hawk in the US, "posing" on a telephone post in Arizona. Photo by George Robertson.

temperature, and during these seasons it does consume less food. It is important to point out that the metabolism of this hawk—as is the case with other hawks—is very slow and tends to store a lot of energy. This is why it is such a slow and difficult process to reduce their weight, especially during the initial stages or when it is "fat" and has not been flown for sometime. However, with a proper diet, a properly trained bird will generally be able to fly well, even on days when they are somewhat overweight.

To capture its prey, the *Buteo jamaicensis* uses various hunting techniques. Being a hawk, it is very opportunistic and may be seen on treetops and telephone poles, maintaining watch there for hours at a time until it catches sight of its prey. At that moment, it approaches its prey silently, gliding down toward its victim, at times beating its wings vigorously while doing so, and frequently hiding from its prey until the latter approaches close enough to be pounced upon and seized without alerting its prey. This method of hunting can be described as "still-hunting", as the red-tail will sit still waiting for its prey and then hunt it when it appears. The red-tail can also hunt at higher altitudes (in moderate or thermal wind conditions). It should not be forgotten that the *Buteo jamaicensis* is generally a hawking bird that can also fly at greater heights just like falcons do, although there are many falconers who still do not appreciate the bird's great hunting versatility. When it sees its prey, it draws in its wings and dives directly toward its victim at great speed (sometimes reaching over 200km/h – 124 mph), falling

The Red-tailed Hawk

*Sequence of attack by a red-tailed hawk on terrestrial prey.
Drawing by Dr. Paul A. Johnsgard.*

Female red-tailed hawk "on guard" gripping its prey, a lure in a flying display, before beginning to eat. Photo by courtesy of J. M. Cabrera.

upon it without warning. Similarly, as is the case with the kestrel, it can appear to hover or fly against the wind, while, inspecting the land below, choosing its prey, and diving toward it, as it does when it sees prairie dogs, for example.

Finally, the red tail also shares a hunting technique that is very similar to that of its larger relative, the ferruginous hawk (Buteo regalis): that of flying at high speed near the surface of the ground, closely inspecting the terrain until it finds its prey.

To attack these prey, the red-tail uses its claws instead of its beak, thrusting its legs toward the prey and initiating contact with one talon in front of the other. It then suddenly grasps its prey, an action which usually results in instant death due to the pressure of the grasp. However, should the prey remain alive, the red-tail will use its beak to deliver the *coup de grace*. During these attacks, and especially when the red-tail is young and inexperienced and can suffer all sorts of tumbles, the nictitating membrane (i.e., "the third eyelid"; see previous photograph and sketch above) of both eyes is drawn over them, protecting the eyes at the moment of attack. Curiously, this is a feature shared by many of the most precise predators, such as funnily enough although not a bird, the feared great white shark as I found out also while studying these creatures.

Despite these protective measures, there are many examples of red-tails that, both in the wild and when used in falconry, present multiple scars and bites on the hind feet, tarsi and head—generally as a result of having been careless or inexperienced. Many of these injuries, when they occur in the context of falconry, can be avoided by using protective chaps, which are widely used in the US, and which will be further discussed in the chapter on equipment.

When an attack is successful and the prey is secured, the *Buteo jamaicensis* begins to eat immediately, assuming a characteristic position (identical to that of the golden eagle), with its hackles sticking straight out. The purpose of this posture is to signify to other birds of prey is that the red-tail is claiming and defending its "meal" against other birds of prey that might otherwise take an interest in it and try to steal it away. Birds of prey typically eat their larger quarry in the very same place that they have captured it. In cases in which they do not consume all of the edible parts at one sitting, they return over a period of days until they have consumed it entirely—providing that it is still in the same place and has not been stolen away. Smaller prey may instead be taken to a tree, where they can be consumed at a more leisurely pace. In falconry, many of these same characteristics are observed, such as protecting the quarry (to prevent it from being stolen, something that frequently happens in nature). This can be observed in the photographs below. In falconry, however, the bird will typically, after securing its prey, look around while maintaining an upright posture, its neck feathers in an upright position—as is the case with eagles—at times seeming to await its master's arrival prior to consuming the quarry.

The Red-tailed Hawk

Typical "eagle-like" posture of a red-tailed hawk, covering its food while it keeps a look-out from time to time to all that is happening around it. Photo courtesy of J.M. Cabrera.

The truth is that red-tails are quite passionate about hunting, and it is at this moment when they will be all worked up, still excited about the chase and hunt, and quite ready to strike out, not even realising that it is us when we approach them. That is why, it is more than recommended with red-tails to approach them slowly and give them a moment to "cool" off, before we make in. I used to even secure my red-tail after that to my glove and just sit beside it, giving it time if it was an excellent catch to feed up on will. This will also be one of the best ways to bond with your bird and gain its trust.

Although it is unusual, on occasions, various red-tails have been observed "cooperating on a single hunt" (Johnsgard, 1990), especially around a tree when trying to trap squirrels. One might compare the behaviour of red-tails in such instances with that of Harris hawks, which are known to frequently hunt in packs like wolves or socially. However, one should make no mistake: red-tails are very independent birds. They are very territorial and will not usually share their prey except in instances where a couple is raising a baby hawk (with the male generally taking the responsibility of providing most of the food to his mate and to the baby hawks). And, even during such times, it is often observed that each member of the couple will often go out and hunt on his or her own. This is something that happens both prior to and following the actual hatching. When such independent hunting occurs, the only sharing that takes place is either on the part of her mother with her young, or in instances when there is actually an excess of food.

The exact nature of red-tail behaviour, in terms of cooperative behaviour, is not entirely known. When they cooperate in capturing squirrels, only one of them eats the squirrel that has been killed and does not share it with the birds who have assisted in its capture—unless there is an excess of food and/or the red-tail who has first partaken of the prey has eaten to the point of satiation. Red-tailed hawks are not especially sociable with one another: they tend to be solitary apart from the times when they form couples—and even then, as we have seen there are limits to the level of cooperation and sharing. In nature preserves, two or more of the species *Buteo jamaicensis* have been observed simultaneously hunting the same mice. In such instances, the birds are clearly not hunting in a cooperative manner. Instead, each bird has its own "spot" or hunting outpost. Such competitive behaviour is only seen when there is an abundance of prey. When prey is scarce, on the other hand, no red-tail will tolerate the presence of another on "its own" hunting grounds.

This kind of behaviour may be observed among northern red tailed hawks after they migrate to the south in the winter months, at which time they enter the hunting grounds of "other" red-tails that are native to those areas. At times, especially when prey are abundant, there is usually not a lot of conflict, although the native red-tails will issue "warnings" to make their presence and property known to the intruders. However, if the red-tailed hawks that are native to the area perceive that their food supply may be in danger, they will not hesitate to confront the invaders.

The migratory hawks typically avoid this kind of conflict and, accepting that they are not in their own territory, tend to act cautiously: at the slightest sign of a problem, they will move to another area. In falconry, this kind of territorial aggression is also observed among red-tails. It is thus very difficult, if not impossible, to get two red-tails to fly together. The cooperative behaviour of Harris hawks stands in stark contrast to that which is observed among members of the species *Buteo jamaicensis*.

A similar situation can be observed among red-tails used in falconry that usually do not tolerate another bird of prey "on their turf". Although the intruding bird in such a case is usually able to return to its own hunting grounds, it at times does so only to find another bird of prey in its garden or on its turf. On one such occasion, the falconer returned to find that the intruding bird had been killed by his red-tail, who had returned to what it considered its "home".

Migration

It is generally true that Buteo jamaicensis is a migratory bird, although there are certain exceptions. Red-tails that live in northern climates, such as the subspecies that reside in Alaska, Canada, and the northern US, are for the most part migratory and generally fly south to spend the winter months. Those that are in southern climates (e.g., Arizona, Texas, Florida, the Bahamas, Jamaica, Puerto Rico and Cuba) usually stay where they are during the winter.

There are two main migratory movements observed among migratory red-tailed hawks: the southward migration in autumn and the return northward in the spring, when they go back to their breeding grounds to reproduce.

Most northern subspecies of Buteo jamaicensis migrate south for periods of 3 to 5 months, although some stay in their northern homes year-round and simply expand their hunting grounds during the winter months. Some of the specimens that do migrate cover 1,000–2,000 miles (1,600–3,200 Km), according to measurements that have been carried out using various "banding" techniques. It has been observed that younger birds are more likely to migrate than older birds and that, among the red-tails that do migrate, younger birds tend to migrate farther than older birds. Many Canadian specimens have been observed to carry out long migrations, only stopping when they reach the Gulf of Mexico. There are some red-tails that will migrate some years and stay put other years. It is not known exactly why this variation occurs, but there are various theories and opinions, the most accepted being that, during years when they do not migrate, the birds have found a zone with abundant prey. Those birds that stay gain the added advantage of less competition, given that the majority of northern red tails do migrate. However there are also other birds of prey that should be taken into account with regards to migration save owls (who in general do not migrate), as not all will migrate and this can also have an effect on the food sources as they may also be competing for similar prey.

The autumn migration sometimes begins for northern red-tails, and especially younger birds who have abandoned the nest and begun to hunt independently, as early as August. It is the younger birds that are particularly anxious to explore and to travel.

For the majority of the species Buteo jamaicensis, however, migration occurs later, in September and October and sometimes even as late as December. The period of highest migration is from the middle October through the first week of November.

According to data obtained courtesy of the Hawk Mountain Sanctuary, 75% of autumn migrations are carried out between October 25th and November 22nd (with November 5–12 being the period of greatest activity). The numerous flights are almost always preceded by meteorological changes in regions north of the Appalachian Mountains.

Migrations may at times be either earlier or later. The precise determinants of migration are not known, but many agree that autumn migration usually takes place before a cold front moves through the area where the birds make their home. According to experts in migration, there are several things that typically coincide with migration: winds, low atmospheric pressure, decrease in

1. The Buteo jamaicensis

(Isla del Socorro)

(Isla de Tres Marias)

■ Migratory

■ Resident

temperature, a rise in barometric pressure, and the presence of a cold front. Another important factor that determines precisely when a migration occurs is the availability of food, since scarcity of food is a sign that winter is approaching, and usually coincides with a cold front. It is at that time that they some of the typical quarry of the animals begin to hibernate, thus reducing the supply of available food. Conversely, if there happen to be an abundance of prey during the period when migration typically takes place, then migration may be postponed.

Birds usually fly toward their winter homes during the day. They fly either alone or with their mates. At times, they will fly in groups, although the red-tail is not a particularly social hawk. Migratory paths tend to follow mountain ranges. Birds coast southward, utilizing their thermal wings. At times, they fly as high as 800 metres (2,625 ft) above ground level, avoiding large bodies of water, and never drifting far from the sight of land.

The migration of the red-tail can be observed in various observatories and other observation points throughout the

The Red-tailed Hawk

US, although the best place to observe red-tail migrations is the **Hawk Mountain Sanctuary** between October 9th and December 1st (with the period of greatest activity being the first week of November). It is estimated that 90% of all migrating red-tails pass through this sanctuary.

In preparing this section on migration, I contacted one of the best places in the world for sightings and research on the migration of birds of prey: Hawk Mountain Sanctuary. The director of conservation projects of the *Acopian Center for Learning* at the facility, Dr. Keith Bildstein, and the rest of the staff, graciously offered me their help and support for this book. The information that follows draws on this invaluable resource.

For the past 74 years, Hawk Mountain Sanctuary has become famous as one of the best places to observe the migration of birds of prey. Some even call the place "one of the wonders of nature," since, from August to December, an average of 20,000 birds of prey, including the red-tailed hawk, eagles, and falcons, glide past the rocky Northern Lookout Point of the sanctuary, 1,500 feet (457m) above Kittatinny Ridge, in eastern Pennsylvania. Many birds even fly at human eye level. Each year, more than 70,000 people go to the sanctuary to be able to observe these birds fly amid the Appalachian Mountains. In 2006, Hawk Mountain counters recorded an astonishing 25,156 hawks, eagles and falcons – the largest number of migrants in two decades.

Migratory flights have been followed continuously ever since the sanctuary was founded in 1934, except for a hiatus of 3 years during the Second World War. The research and census of birds that has taken place over the past 74 years represents the largest record of populations of birds of prey in the world, and contributes much valuable information about the change in populations of these birds in the north-eastern US and Canada.

This migration is important for biologists, since it represents a unique opportunity to estimate the populations of these birds of prey. Currently, the annual Hawk Mountain census allows scientists to both determine long-term and short-term behavioural patterns of migratory birds of prey over the course of decades, and to evaluate the health and environment of each species.

Hawk Mountain also carries out studies on the migratory behaviour of birds of prey, its effects on local fauna, and long term behaviour patterns of populations of birds of prey. Its new field study and research centre, *The Acopian Center for Conservation Learning*, has allowed conservation efforts to expand. This research centre has offices, a library, a laboratory, and storage rooms with maps and archives. In addition, it has equipment and materials for visiting scientists, residents and students who are doing field work.

The phenomenon of birds of prey following the Appalachian Mountains southward has been occurring for thousands of years. Weather conditions may affect the number of birds that are observed on any given day, with the greatest number of birds observed following

1. The Buteo jamaicensis

Two images of the North Lookout at Hawk Mount Sanctuary where, every year, we can see many amateurs and experts spending the day observing and collaborating with the staff of the facility, where autumn migrations have been documented since 1934. When birds begin to migrate southward from Canada, New England, and New York to flee the winter climate and escape the scarcity of food, the staff and many of the visitors gather together to count all of the sightings. Images courtesy of Dr. Keith Bildstein, Director of Conservation of the Acopian Center for Learning, Hawk Mountain Sanctuary, Pennsylvania (USA).

The Red-tailed Hawk

passage of a cold front through the area. This is one of the reasons that the mass migrations observed on a given day may not be seen on that same day the following year.

As for the manner in which migration usually occurs, birds of prey, including the Buteo jamaicensis, carry out their migratory flights during the day, using so called "thermal" air pockets to travel long distances while saving energy. Thermal columns allow birds of prey to glide along the mountains, saving energy during their southward journey. Because thermal columns do not occur over the water, birds of prey do not stray far from the Appalachian Mountains as they fly toward their winter homes.

Rather than flap their wings, raptors conserve energy during migration by soaring. There are two types of soaring: thermal and slope. Thermal soaring takes place when a raptor circles up and within a rising warm-air tunnel caused when the sun differentially heats the surface. Slope soaring occurs when raptors 'hitch' a ride on winds that are deflected upward by hills and mountains. The following drawing, which was done by Hawk Mountain, illustrates this process.

Some birds, such as the red-tailed hawk, carry out migrations of thousands of miles toward their warmer winter homes in Central America. Some specimens of the Swainson's Hawk migrate to locations in South America. Others, such as the Bald Eagle, spend their summers in the south-eastern US. Some species are rarely observed during migrations. One of these is the Mississippi Kite, which was observed at Hawk Mountain on October 8, 2002. Another was the Long-eared Owl, observed on November 11th of the same year.

Migrations of singing birds and insects, including butterflies and aquatic birds, have also been recorded at Hawk Mountain.

On the following page we can find a table of data compiled at **Hawk Mountain Sanctuary** based on research carried out during the past 60 years. It contains the dates that are most conducive to observing particular subspecies of hawk during the day.

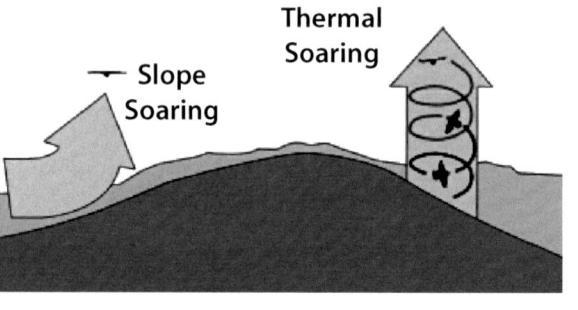

As we can see in the table, the period from mid-October (at which time the greatest diversity of species can be observed—about 16 different species in total) until the beginning of November, is the best time for observing migratory red-tailed hawks. Some specimens of red-tails, however—especially those living farther north, may begin their autumn migration before the others, beginning as soon as August in some cases. The Hawk Mountain observation program officially begins on August 15th each year.

1. The Buteo jamaicensis

Species	Aug 15–30	Sept 1–14	Sept 15–30	Oct 1–14	Oct 15–31	Nov 1–14	Nov 15–30	Dec 1–14
Osprey	57	87	89	73	29	4	1	<1
Bald Eagle	40	52	30	14	10	15	18	17
Northern Harrier	49	76	79	83	86	77	42	21
Sharp-shinned Hawk	41	84	98	98	96	79	41	11
Cooper's Hawk	12	32	66	85	72	45	17	13
Goshawk	2	2	7	20	36	46	50	33
Red-shouldered Hawk	3	6	18	63	83	73	45	15
Broad-winged Hawk	87	94	94	42	4	<1	<<1	0
Red-tailed Hawk	39	51	66	88	96	96	88	79
Rough-legged Hawk	0	0	0	3	12	21	18	21
Golden Eagle	0	3	10	27	45	50	42	26
American Kestrel	68	77	77	74	40	8	1	0
Merlin	3	8	20	42	28	5	1	<1
Peregrine Falcon	4	10	22	44	19	6	3	<1
Poss. species	12	13	13	14	14	14	14	12

Table for autumn migration, adapted and published with permission of Hawk Mountain Sanctuary for this book, based on sixty years of data.

As the table shows, the variety of species and the number of specimens observed vary from one day to the next, depending on weather conditions and the season of the year. The autumn migration usually begins in the middle of August with small numbers of various species, including American kestrels, ospreys and bald eagles, as well as small numbers of red-tails from the farther reaches of the north. By the middle of September, one begins to observe larger groups, or "kettles" of broad-winged hawks. Such groups may consist of as many as 1,000 individual birds.

The flights of the majority of specimens and species usually occur in the middle of October, at which time one can observe royal eagles, red-shouldered hawks, and numerous red-tailed hawks, among other species. From that

The Red-tailed Hawk

time onward, the numbers of birds observed decrease until the end of the autumn migration in mid-December, although eagles, goshawks, and rough-legged hawks have been observed migrating even as late as January. In the spring, one can observe the spring migration at Hawk Mountain as well, although fewer birds pass through the sanctuary in the spring, because the Eastern winds generally push birds west of the site. The greatest number of specimens during the spring migration can be observed during the third week of April. However, some red-tails—specifically, those residing in areas in the far north, may begin their spring migration as late as June.

	Total Year 2002	Average 5 years
Black Vulture	8	4
Turkey Vulture	8	19
Osprey	129	114
Bald Eagle	10	5
Northern Harrier	32	28
Sharp-shinned Hawk	114	132
Cooper's Hawk	67	28
Goshawk	0	0
Red-shouldered Hawk	4	8
Broad-winged Hawk	266	387
Red-tailed Hawk	**108**	**107**
Rough-legged Hawk	0	0
Golden Eagle	1	1
American Kestrel	36	42
Merlin	6	5
Peregrine Falcon	0	1
Unidentified accipiters	14	17
Unidentified buteos	27	29
Unidentified eagles	1	1
Unidentified falcons	3	2
Unidentified raptors	27	45
Total	**861**	**976**

Table of raptor spring migrations of the last few years, adapted and published with permission from Hawk Mountain Sanctuary for this book.

1. The Buteo jamaicensis

It is interesting to note that, according to the table below (which was supplied by Hawk Mountain), the last migratory specimen observed during the 2002 spring migration was a solitary red-tailed haw that was gliding along the North Ridge of the mountain on the 15th of May.

Hawk Mountain has two programs for observing migrations: the autumn migration (August 15th-December 15th) and the spring migration (April 1st-May 15th).

Hawk Mountain collaborates with amongst others the National Audubon Society and the Cornell University Ornithology Laboratory to coordinate "Hawk Watch" each fall, using the Internet to link various observation posts throughout the Americas. Hawk Mountain invites established observers to participate in this observation network that extends from Canada through Central America. Participants simply need to record their sightings on the network. This website obtains, analyses, and shows the results of the migration of birds of prey using maps and tables in real time. The website also uses information about the number of birds of prey in order to control bird populations. These data are published by the North American Association of Migration of Birds of Prey.

These programs of observations are very important for migratory studies of populations of birds of prey, and they are frequently used together with data obtained from other observatories in the US, and in other nations.

Other important observatories in North America for sightings of birds of prey in both autumn and spring are:

Autumn migration
- Cape May (New Jersey),
- Cape Charles (Virginia)
- Goshute Mountains (Nevada)

Golden Gate Raptor Observatory (California)

Spring migration
- Derby Hill Bird Observatory (New York)
- Braddock Bay Raptor Research (New York)
- Whitefish Point (Michigan)
- Dinosaur Ridge Raptor Migration Station (Denver)

For more information on Hawk Mountain and other raptor conservation and observation centres, please refer to the useful addresses section at the end of this book and visit their website at **www.hawkmountain.org**.

"Banding"

Photos and article by Brad. S. Silfies

One of the best ways to capture and study the largest number of birds of prey is to make use of a banding station in a place where there is a high concentration of migratory movements.

Coastal areas, mountain ranges, peninsulas, or bottlenecks in large bodies of water can attract a much larger number of birds of prey than normal during the autumn and spring. Think, for example, of how difficult it would be for someone studying the plumage of birds of prey to be able to capture a significant number of specimens simply by visiting the wide open areas that these birds typically inhabit. The fall represents a unique opportunity for researchers to see how well species have done. By means of consistent efforts, year after year, these studies can assist those responsible for making decisions with respect to the control of birds of prey. Banding stations remain in the same place that they were initially established in order to maximize their exposure to the sky. An effective means of attracting birds of prey in flight is by setting out some food for them.

The capture of the majority of birds of prey requires some type of moving "bait or lure" in order to alert the bird of the presence of a potential "meal." Pigeons are often used for bait because they are tough birds and with a pigeon harness are the ideal lures for migrating red-tails. The bait that is

Banding station in Little Gap, PA, USA.

1. The Buteo jamaicensis

used is connected to a rope that is anchored in a hiding place—where the bander is located—and is placed on a pole that is at least six feet high (1.83m). The pigeon is pushed up into the air, and then begins to flap its wings vigorously. In this manner, it may attract the attention of any red-tail that is nearby. Near the hiding place, the bander has place a spring-loaded bownet that is triggered if a bird approaches the food. The pigeon frequently flies upward so that birds of prey can see it from far away. Once the red-tail has committed itself to the attack, the bander pulls the pigeon toward the centre of the bownet trap and, if the raptor lands nearby and closes its wings, the bander activates the trap. He will then run toward the trap to secure the bird, and then reset the trap. The most dangerous aspect of managing any bird of prey is finding its feet.

People generally think that the beak is the most dangerous feature of any bird of prey. I think that this is because, in the majority of photos that one sees, both the head and the sharp and powerful beak of the bird are prominently featured. Very few birds of prey actually bite (with falcons being an exception). The best way of controlling the feet of a bird of prey is to put the bird on its back and to have its feet facing you. In this way, you will be able to see exactly where the feet are, and this is an important advantage. Generally, you will find the birds feet extended toward you, pushing against the net. This is another advantage, since the bird will have to retract its feet toward itself in order to attempt to strike you. These birds are very fast and, for this reason, the best option is to distract them. Hold your hand above your head. The bird

Close-up of one of the characteristically strong, powerful feet of the red-tailed hawk.

The Red-tailed Hawk

will be distracted and will look toward it. Then quickly grasp the bird's two feet with your free hand. This should be fairly easy, as the birds feet should be somewhat relaxed. I do not recommend the use of sticks, towels, gloves, or anything else that could injure the bird or damage its plumage. One other piece of advice is to keep everyone away until the bird is secured.

It is a good idea to take some pictures at this point, since the bird will have its wings extended or open, and the feathers of its neck will be standing on end—given that the bird I son the defensive in the face of this threatening situation. Later, you will actually have to spread its wings in order to be able to take a photo of the bird with its wings open. After resetting the trap and checking the bait to make sure that it has not been injured, the captured bird of prey is taken to the hiding place. Birds should be carried by cradling them with both arms so that their wings don't move (and, thus, to reduce the possibility of injury to the wings).

The bird should then be visually examined, and its measurements should be taken. Its tarsus needs to be measured with a special gauge in order to determine the size of ring that will be use to band it. Because there is so much variation among red-tails, this ring may be one of three different sizes that are typically

A bownet set to capture the next red-tail, as part of a "banding" program.

used. We have captured birds ranging in weight from 695 g to 1,500 g (24.5–53 oz). For red-tails, closed rings that have a flap that have a backward folding tab are used. These rings are impossible to remove or to pull off. Once the ring has been placed on the bird, the bander needs to make sure that it can slide along the ankle of the tarsus where it has been placed, and that it does not in any way impede normal movement. The first thing that is usually done with birds of prey that are captured in banding programs is to place rings on them, so as to avoid the possibility of someone forgetting to do this afterward.

The number of the ring is recorded in a field guide, as are the bird's physical measurements. Then the distance from the tip of the longest primary feather to the curve of the wing is measured, with this measurement constituting the arc of the wing. The metric system is used for all measurements. The width of the wing with the second-longest primary feather is recorded as a means of determining the sex of the bird. It is not possible to determine the sex of red-tails with absolute certainty, but it is known that larger specimens are usually females and that smaller specimens are usually males. One study carried out on 100 birds whose sex was definitively determined after they were measured revealed a great deal of overlapping among the sexes with respect to size. For this reason, it is really only those birds that are extremely large or extremely small whose sex can be determined with a high degree of certainty.

The tail is measured by placing a ruler in between the two central feathers of the tail and pushing it toward the body until it

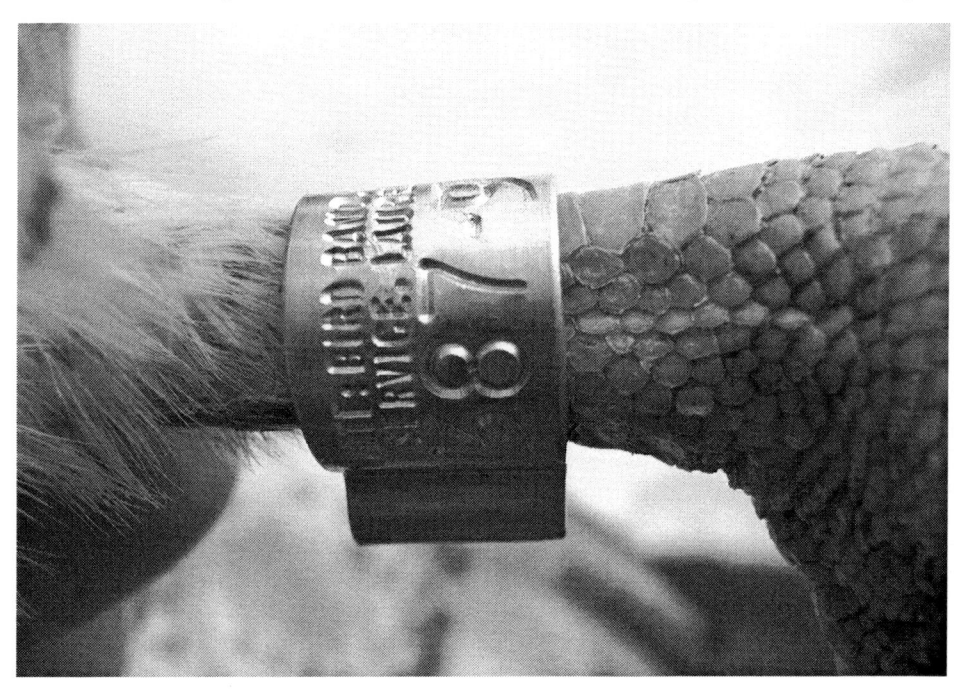

A red-tail that has been ringed as part of a banding program.

The Red-tailed Hawk

can no longer be moved. This will yield a measurement of the longest feather of the tail. Juvenile and immature adult specimens generally have longer and narrower tails than mature adult specimens. The tail of juveniles is generally brown, and contains 7 to 9 dark bands. The tail does not acquire its characteristic red colour until the age of two years. Many birds retain one or two brown feathers along with the red feathers of the tail during the following fall.

The bander takes careful note of the bird's plumage during the banding process, since any feather that appears to be moulting can aid in determining the age of the specimen. A diagram showing the feathers of the bird, and that indicates which feathers are missing, broken, or folded, as well as the new feathers that are emerging, can aid in determining the age of the bird. The wing feathers, both primary and secondary, can help us determine if a specimen is two years or older, which is highly probable if there is a greater distance than normal between the new and the older feathers.

However, the tail is the best indicator of age of the bird, and whether or not the bird has experienced its first change of coat. A tail that is uniformly brown in colour indicates a bird that is one year old or younger, and is marked as "HY" (for Hatching Year"). A tail that is completely red in colour belongs to an adult and is older than one year, and is labelled "AHY" (for "After Hatching Year"). A red-tailed hawk that has at least one brown feather is labelled "SY", indicating that it is two years old.

Juvenile specimen of the Eastern subspecies of the red-tailed hawk, which has just been captured by Brad and which strongly resembles Bandit when he was a chick.

1. The Buteo jamaicensis

The "culmen" is also measured: this is the distance between the tip of the beak to the beginning of the cere. We measured the longest claws and recorded the measurements. The measurements for the length of the body and for the wing span are only carried out if there are two people who can do the measurements. Red-tailed hawks usually have a wing span of between 1,000 and 1,300 mm (39–51 in). These measurements give us a general idea of the size of the bird.

The colour of the eyes, the cere and the feet are noted, along with any injury on the fleshy areas of the birds that may have occurred prior to "banding". By this time, the red-tail has probably calmed down considerably: it may have relaxed its wings and the feathers on its neck may no longer be standing on end. The term "hackles" is used to indicate the feathers on the upper back portion of a bird's head: It is these feathers that a bird can raise in order to appear larger and stronger.

The bird can now be weighed. The wings of the bird should be folded across one another so that they are kept closed. The bird should then be inserted into a rigid plastic tube so that it cannot move when it is placed on the scale. This tube keeps the bird calm while it is weighed, and thus enables the bander to obtain an accurate weight. Once the necessary information has been recorded, the bird can be taken out of the tube. It is necessary to look into the tube to locate the feet. At this point, it is important to secure the bird

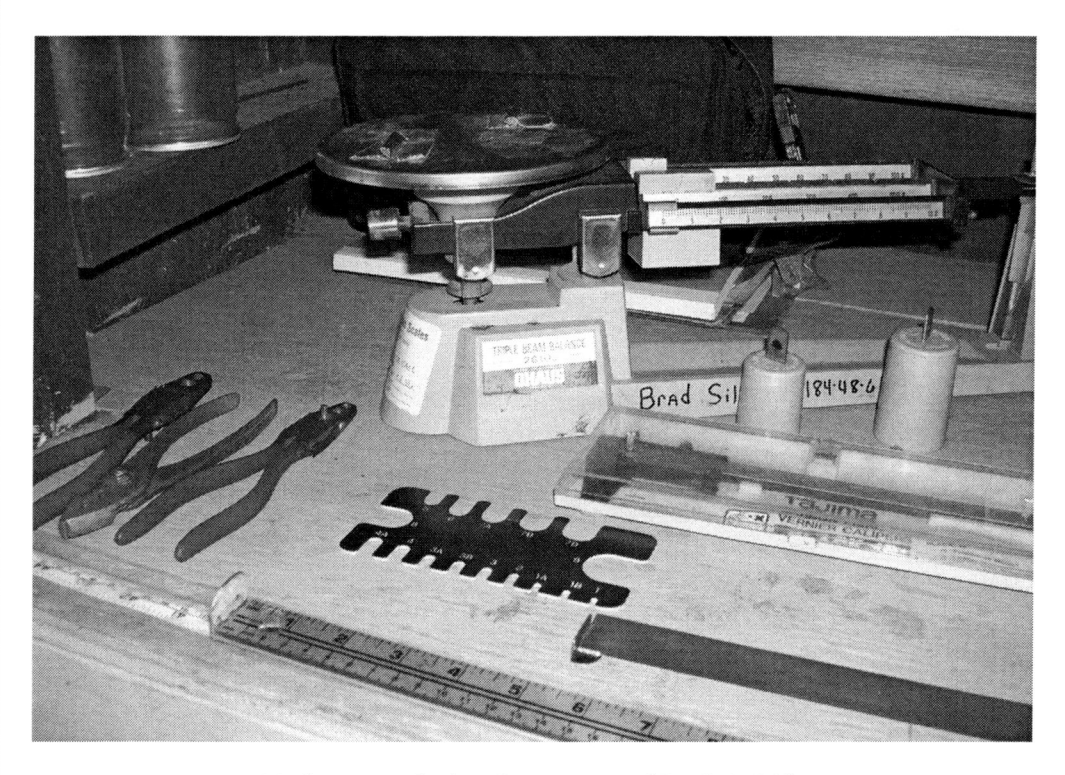

Work area in the banding station of Brad. S. Silfies

in order to avoid serious injury, and to carefully remove it from the tube. Once it is out of the tube, the bird will be bating, opening its wings and with its hackles raised once again.

Normally, this is the time when I determine the proportion of fat that the bird has. By blowing softly on the undersides of the wings, near the points where the wings join the body, as well as on the fulcrum, on the upper part of the chest, we can expose the skin and see if there are fat deposits. The skin normally has a pinkish hue, but if there are fat deposits, the skin will instead be yellow or white in colour. A scale of 0 to 3 is used to indicate the quantity of fat that the bird possesses. The bird is then taken outside and, if more photos are needed, they are taken at this time prior to releasing the bird.

Banding is rather similar in practice to fishing and hunting in many ways. It is a game between man and animal in which one has to sit and wait, trying to coax an animal to come close enough to that it can be caught. Banding programs capture and release their prey—as is also currently the practice in many fishing areas. The more information one has on the preferred habitats and prey, its migration periods, and its daily habits, the more one is equipped to study the species in question—in the present case, the red-tailed hawk. In hunting and fishing, one must act in accordance with the law and have the proper license. This is also the case with respect to banding programs in the United States.

In the United States, permits at both the state and federal levels are required in order to carry out banding. The best means of obtaining these permits is to apply for them after volunteering at a banding station for several years. You should also apply through a person who already has the required credentials and who can verify your knowledge and experience. The majority of professionals who work in banding programs have learned from other professional banders who have more experience. The best way to learn to capture, handle, tag, and set free birds is to volunteer in a banding station. There are many people who devote a great deal of their free time to banding birds, and there are professional ornithologists who band many specimens in their spare time, and without receiving compensation, at field stations.

Banding programs are a great way to introduce the public to the treasures of wildlife that surround us, but who live in areas beyond our reach. Public displays and exhibitions on weekends (or special exhibitions for students) are also excellent ways of involving people in providing support to the banding station. The free time that is donated is a wonderful gift to the area's wildlife, and a well-conceived banding program is worth the effort required. And whenever visitors look into the eyes of a bird of prey from just a few metres (about 6–8ft) away, before the bird is set free, and when they see the beauty and the untamed spirit of this bird of prey, they will remember that moment, and they will want to experience it again.

Reproduction

The Buteo jamaicensis is monogamous, partnering with a member of the opposite sex for life. However there have been observed rare exceptions of nests with two females and one male. But, generally speaking, the species forms dyads for the purposes of mating, and partners are only changed when one of the birds in the couple dies. When there is a death, the surviving partner will typically act quickly to find a replacement partner. When the male dies, the female normally stays in the same territory and will attract another male. But when the female dies, the male generally needs to leave his territory in order to look for a female in another territory.

The red-tailed hawk is generally not considered capable of reproducing until it obtains its red tail, which usually occurs during its second autumn of life, when it is more than one year old and its first moulting is complete. The birds have usually finished growing by the time they have reached the age of two. However, there have been cases reported of red-tails copulating prior to acquiring their red tails—thus showing evidence of sexual maturity even before they have shed their juvenile plumage. It seems that there are fewer females than males. It is also possible that the females are simply more difficult to locate. Whatever the case may be, females seem to be much more manageable in a falconry husbandry context than males during their initial year of courtship.

With regard to determining the sex of red-tails, it is important to keep in mind that, even though females of the species are usually larger and heavier—a difference seen among many other birds of prey as well—size and weight alone cannot be used to determine sex. In fact, in the United States, where many individual specimens of red-tails and other birds of prey are followed by means of "banding," these programs generally do not allow the bird's sex to be determined on the basis of size alone. The only way to be sure of a bird's sex is via laboratory or veterinary examination. This is because there have been cases reported of large specimens that appear to be females, but that turn out to be males, and *vice versa*.

According to some experts in the subject, sex appears to be related to such variables as thickness of the head, curvature of the beak, and size of the tarsi and claws. There are females that are very small and males that are very large, but even the small females generally have broader shoulders, while large males tend to appear as less stocky, irrespective of their weight.

There are also mathematical formulas that can be used, with varying degrees of accuracy, to determine a bird's sex that take into account various body proportions and weights but which will require the results to be compared against the feathers of live individuals. Unfortunately, all of these tests do tend to have a have a very high margin of error, so this is not the most recommended method for sexing.

When I wrote this book about 4–5 years ago, the most frequently employed method in the US for determining the sex of raptors was DNA sexing; it yielded the most accurate results and was made up of various blood tests that required a minimum amount of blood in order to determine the sex of an individual specimen. The tests were sold in "kits" for various different birds of prey by some specialized laboratories, and could be easily carried out at home. This blood is then sent to a laboratory and the results are can be available the very next day.

This still continues to be the most effective method for sexing birds of prey that we have available today, and is the one currently favoured by zoologists, breeders, and falconers. It has an error rate of only 0.01. The test on a single blood sample is repeated several times in the laboratory to assure that an accurate result is obtained. This procedure yields far more accurate results than the analysis of feathers which was the standard procedure for many countries in southern Europe until a few years ago, and since then DNA testing has increased in popularity. Many breeders actually now provide DNA certificates with their birds and if not, you may ask for it to make sure you get what you pay for when acquiring a new bird as the tests can determine the sex of many of the raptor species commonly used in falconry.

Returning to the subject of reproduction, the formation of couples, or courtship, among members of the species Buteo jamaicensis usually begins at the end of winter, and frequently occurs far from the zones where they mate—i.e., in the lands where they spend their winters. During these times, one can see the red-tail couples carrying out acrobatic manoeuvres, including soaring in wide circles at altitudes that are so high that they are difficult to observe. The male will perform a series of spectacular dives in front of the female, with the two of them coming into contact and binding into each other letting their legs dangle in the air between each manoeuvre.

Later, the male will try to approach and come to a position just above the female, touching her with his feet extended. Sometimes, the male and female touch one another with their beaks and claws. When the two birds hold each other in one another's claws, they fall in a spiral toward the ground at great speed. The birds typically disengage from one another before landing on the ground. However, incidents have been observed in which the birds have fallen to the ground at high impact, bouncing off the ground before disengaging and again taking flight, apparently none the worse for wear. These manoeuvres usually last five to ten minutes and are usually accompanied by repetitive calls by one or both red-tails—calls that each last a few seconds.

During the period of courtship, one can frequently observe the male hunting prey and then offering it to the female. This is important, since the female may see this as an indication of the hunting abilities and health of the male. It often will be the male that is responsible for providing food for the female during the time that the eggs are incubating, and then later after they hatch. Sometimes, however, the male and female will share this responsibility during this period. This courtship ritual may also serve the function of assuring

1. The Buteo jamaicensis

Close-up of a juvenile specimen of an eastern red-tail. In this photo, we can observe two prominent features of this buteo: its aggressiveness and its nobility. These two characteristics serve to remind us of the respect that we should have toward all of nature's creatures. Photo by Brad Silfies.

Here we can see Bandit's tail after the first moult with some feathers not fully grown still. As an eastern subspecies, we can see that the tail is a light red hue, without any intermediate black bands or stripes save for the black subterminal band, a little narrower than in other subspecies and its point ending in a "spike". Photo by author.

The Red-tailed Hawk

Bandit's juvenile tail, before his first moult and after an exciting chase of a pigeon plunging right through all sorts of bushes and conifers, basically just ploughing right through anything that got in the way, including the author! If we look closely at the tail we can see the typical barring of the tail in juvenile red-tailed hawks, and the tip of the tail ending in white, a trait common to eastern subspecies. Photo by the author.

In this image, we can see a breeding pair of red-tailed hawks, male and female, and it is very easy to tell them apart, due to the obvious difference in size. However, size as well as other features such as the fact that females are more heavily marked (in this case it is the opposite, the female is actually lighter and with hardly any markings), will not be sufficient to determine the sex. It will be necessary to test the individuals through a blood or feather test, the first one being much more accurate and highly recommended. Photo by the author.

that the female receives the nutrition she needs to produce eggs and begin the process of raising her young. The availability of food is important at the time of reproduction, since it has been observed that in places or at times when food is scarce, fewer eggs tend to be laid.

This courtship behaviour may be observed not only during the breeding period, prior to the eggs being laid, but also, sporadically, at other times. This behaviour thus seems to reinforce the bond between the couple.

Red tailed hawks are generally very territorial, especially during the breeding period: They aggressively defend their territories, which are defined by both members of the couple. The territory and its size may depend on the availability of prey, the characteristics of the terrain, and the presence of competitors. Some couples remain in the territory year round, but may go to hunt in other zones that our considered "foreign territory". These territorial zones of the red-tails are defended against both other red-tails and other species of birds of prey. In addition, these territories are defended against human intruders. There have been more than a few reported instances of attacks by red-tails against human beings, generally when the latter have unknowingly approached a nest of red-tails. It should also be said that, even though red-tails can adapt to the presence of humans and can grow accustomed to all kinds of noises, they are by nature very shy, especially as adults, and there have been a good many instances in which they have abandoned their nests instead of attacking intruders, leaving eggs behind that never hatch. For this reason, it is strongly advised not to disturb red-tails, if at all possible. The same applies, indeed, to all birds of prey during the time of breeding, since any disturbance might result in an unfortunate incident that might have been avoided—and perhaps provoked in the first place only to get a photo or to see a nest. If you do decide to breed red-tails in captivity, please consider where you build their breeding chamber and choose the quietest spot.

In the wild, the nest is usually selected by both members of the couple, who afterward return to their former nests to tidy them up for many years afterward. The nests chosen by the newly formed couple are generally in secure locations, usually very high above ground level. This is because the species generally prefers places that are both inaccessible and that afford a good view of the land below—a feature that facilitates the couples' control of the zone. Nests constructed in trees—frequently beech, maple and pine trees, are often near a clearing, where they will be able to hunt. Red-tails have a tremendous ability to adapt, and often will utilize any solid structure to make a nest.

In desert areas (e.g., the south-western US), and Mexico they will often, like Harris hawks, construct their nests on saguaro cacti. In other zones like the north-eastern US, red-tails will utilize, in addition to trees, inaccessible cliffs and rocky areas. Red-tail nests may also be seen in urban areas, such as metropolitan New York City, where for many years many red-tail couples have been observed in Central Park, some constructing their nests on buildings along Fifth Avenue near the park, where they feed mainly on pigeons and small mammals. Red tails have found urban homes in other US cities as well, although these are exceptional, rather than typical. In some places, red-tails have built nests behind billboards, on telephone poles, and atop electrical towers. But, again, such sites are atypical, given the species' avoidance of any human presence.

The time when nests are constructed and territories established varies among different zones and subspecies of the red-tailed hawk, but generally occurs around the end of February and the beginning of March. Red-tails have been observed beginning construction of their nests long before this period of time, some as early as December (in some parts of the southern US, like Arizona) with others beginning the construction of their nests as late as April—or even later, in rare instances.

From what I've learned, it is generally the sedentary (i.e., non-migratory) birds residing in southern climates that begin constructing their nests earlier, and the migratory birds native to northern climates that begin at a later date.

The nest is made of twigs and branches that are a couple of centimetres in diameter (0.4–0.8 in), with green conifer twigs which are placed on the outside of the nests at the beginning of construction. Some experts believe that this is intended as a sign that the nest is under construction. They also seem to serve the purpose of acting as natural pesticides, avoiding insects and parasites.

The diameter of a typical red-tail nest is approximately 63–76 cm (25–30 in) and the depth of the nest is usually 13–15 cm (5–6 in). the bottom surface of the nest is covered with bark, green conifer twigs, and other green material. Both the male and female are involved in constructing the nest, an activity which usually takes between four and seven days. But it is the female who gives the nest its shape and who lines the bottom surface. Nest construction usually takes place during morning hours, with the bird or birds keeping an eye out for any intruders that might approach. If too many interruptions occur during the process of nest construction, then the nest will be abandoned for a different location.

The nests that are constructed are generally used over a period of many years, being periodically refurbished with more twigs for its walls and more bark and plant material for its surface. It is often the case that, in the case of nests that are already built, birds will build several alternative nests in the same area before deciding on which one the female will use to lay her eggs. This process involves the red-tails "trying out" the different nests, not only sitting in them but also acting as if they are protecting hatched eggs. In this way, the couple arrives at a final decision as to which nest they will finally use. At other times, after having used and refurbished the same nest year after year, they might suddenly build another one on a different site, where they might raise young for one or more years before returning to their original nest.

It is interesting to note that the nests constructed by red tails are later used by members of other species as well, including great-horned owls, red-shouldered hawks, broad-winged hawks, the Swainson's hawks, ferruginous hawks, and crows. The reverse also sometimes happens: i.e., some red-tails decide to appropriate the nests of other birds of prey for their own use. In both instances, nests are often reused several times within a single year by red-tails and other birds of prey and this is common in nature. The varying nesting times of different birds allow this to happen: for instance, the great-horned owl nests before the red-tail, from early November[1] through the end of January thus abandoning its nest prior to the time that the red-tail begins the process of nesting). There have been instances observed, however, of red-tails and great horned owls nesting near one another at the same time.

As is the case with other birds of prey, including the ferruginous hawk and the golden eagle, nests of the *Buteo jamaicensis* may also have other "lodgers", such as sparrows, that seem to carry on a mutually beneficial relationship with the red-tails, feeding on insects found in the nest— and in this way keeping the nest clean—while at the same time receiving protection from the red-tails against other predators.

As many as five weeks elapse from the time that the nest is selected and constructed until the female lays her eggs. Among northern migratory birds, however, the period is often shorter—generally no longer than three weeks. This is because there tends to be a shorter period of time during which conditions for incubation are optimal.

The female red-tail generally lays between one and four eggs during a single breeding period. This follows the general pattern for large birds of prey. Most frequently, the number of eggs is either two or three, although there have been reports of as many as five. As a general rule, western red-tails lay more eggs than eastern red tails.

Normally, the female *Buteo jamaicensis* lays eggs once a year. If the eggs from this first set are destroyed, she will lay eggs a second time. It is rare for a third or fourth set of eggs to be laid within a single year, although there have been reports of such occurrences. Third and fourth sets of eggs are usually laid in nests that are different from the primary nest. The colour of the eggs is off-white with reddish-brown specks, with some eggs being completely covered by such specks and others having the specks only on certain parts of the egg. Red-tail eggs are similar in size to chicken eggs (i.e., 48.3–61mm / 1.9–2.4 in).

The female lays one egg every one or two days until all eggs have been laid. Both the male and female of the couple often share duties with regard to perching atop the eggs (i.e., incubation) while each hunting for themselves. This incubation is crucial in order to maintain the minimal temperature necessary to sustain the growth of the developing embryo (i.e., 37.8–38°C / 100–100.4 °F). In many cases, however, it is the female who does most of the incubating,

---

[1] R. Austing. The World of the Red-tailed Hawk.

The Red-tailed Hawk

while the male assures that food is provided for the couple. If both parents incubate, the male will spend less and less time doing so as the time for hatching approaches. This is very important, because if the eggs are outside of the nest for too long, they are in danger of not being properly incubated, and therefore of not hatching. During the period of incubation, the female turns the eggs over every so often with her beak in order to avoid the embryo sticking to the eggshell which would cause problems when hatching, and to assure that the eggs are properly incubated.

The period of incubation usually lasts 28–35 days, and as the hatching day draws near, the chick begins to chirp while still within the shell of the egg. This chirp is a call for the mother, and the mother responds to the call. When breeding red-tails and waiting for some eggs to hatch after incubating them, I was surprised to find that you could actually hear the chirping. The chick begins to break the interior of the egg shell with the tooth on the upper part of its beak. This tooth falls out after the chick breaks out of the egg. Chicks sometimes take an entire day to make the initial hole in the egg shell. Afterwards, they may take various hours to emerge from the shell. Once they do so, which may take various days, given that incubation begins with the laying of the first egg, the chicks are very weak, and white in colour. They have heads that are very large in proportion to the rest of their bodies, and they are unable to lift their heads, remaining nearly immobile during the first hours following hatching.

The birth weight of red-tail chicks is approximately 60g (2.12 oz), with no sex difference observed with respect to size, weight, or colour. Differences between male and female chicks become observable only after about 29 days.

Great-horned Owl "squatting" in a red-tail nest in winter, just before the red-tails move in during spring for their breeding season. Photo courtesy of Geoff Dennis.

1. The Buteo jamaicensis

Captive red-tail breeding pair of Ed Hopkins, United Kingdom, incubating and looking after their clutch. Being a western pair, the usual clutch will be of 3–4 eggs, as we can see in the picture, and very similar in appearance to the eggs of the European common buzzard. Both parents will incubate, though sometimes, during the last few days, the female will do most of the incubation. If red-tailed hawks are not bothered too much during breeding season, as they do not take kindly to the presence of man, can be the best of parents. Photo courtesy of Ed Hopkins.

Female red-tailed hawk in Ed Hopkins' breeding centre with the first red-tailed chicks of the year, only a few days old. Photo courtesy of Ed Hopkins.

The Red-tailed Hawk

Female red-tailed hawk with her 4 week-old chicks in the nest. Photo courtesy of Ed Hopkins.

Liz Hopkins with her male red-tail "Mavrik". Photo courtesy of Ed Hopkins.

During this period of incubation and, especially, following the hatching of the chicks, it is usually the male that hunts for food and provides meals to the female. The male will either bring the food into the nest or leave it on a nearby branch, so that she does not have to go very far from the nest in order to retrieve it. She will then tear the food into little bits what has been left in order to feed her chicks. On some occasions, however, the male and female continue to share hunting and incubation duties even following hatching. However, it is always the female who breaks down the food and feeds it to her young, while the male either stays near the nest watching or flies off to hunt more prey. There is a theory to this regard that has been greatly debated and proposed as an explanation for the difference in size. It justifies the larger size and bigger intake of calories needed by the females as they need more energy to feed and rear the chicks against the smaller proportions of the males, required for greater speed and agility when hunting.

As for the prey that are most commonly given to chicks for food, these are the identical to those hunted and eaten by adults, except they are broken down into very small pieces first, and the inedible portions of the dead animal are then removed from the nest. The chicks of Buteo jamaicensis are fed by the female from the time they are born until the oldest of the chicks is 30 to 35 days old.

On some occasions, two females and one male have been observed sharing a nest with four chicks (Wiley). This type of arrangement has been termed "cooperative raising" by some. In such instances, the male provides most of the food, with the females later breaking down the food and feeding the chicks, protecting them from the elements, and defending the nest.

The chicks grow at a very rapid pace. On their second day following hatching, they are already active and chirp softly for their mothers. At one week, they begin to take an interest in the prey brought into the nest to feed them, and begin emitting piercing chirps. At two weeks, they typically have begun breaking down their prey themselves. At three weeks, they begin spreading their wings and, by thirty days following hatching, are exercising their wings on a daily basis. Chicks typically leave the nest after about 45 days. When they leave the nest, they remain close by for several days, often following their parents as if expecting to obtain food from them. The parents will continue to provide food for their young for the first three to four weeks following abandonment of the nest. These young red-tail chicks will begin to capture vertebrate prey at the age of about six weeks. At this point, they begin fending for themselves, generally leaving the zone where they were raised in search of their own private "zone", often choosing for this purpose an area where there are a lot of other young red-tails. The young red-tails are often very playful, using their beaks to toss sticks and twigs to one another. They will often head south earlier than adult red-tails, anxious to explore the wide world that lies before them.

Breeding Red-tails

Article and photos by Paul Beecroft

1998 was my first attempt at breeding red-tails. I paired them up in January and from the start there were no problems in the aviary. The aviary was in the main skylight and seclusion except for a drop down trap door approx. 15" x 6" (38 cm x 15) which to begin with was left open. They commenced calling in late February and early March and at that time I started putting nesting material into the aviary and also closed the trap door. They commenced building immediately and as fast as I put it in it was being used.

On April 17th the female was sitting and I assumed that she had laid her first egg. This was confirmed some three days later when I was able to see two eggs in the nest. Over the following weeks I was only able to view them through a peep hole but during this time I was able to see that quite often the male would be incubating on his own with the female perched close by. On a number of occasions I also saw both of them in the nest together side by side.

Now, as I am sure we have all experienced the hatch time can come at the most inopportune moments as far as us humans are concerned and in my case it clashed with the Falconers Fair. The problem was solved however with my wife Lynn volunteering to stay at home in case of any problems. When I left on the morning of 24th May, although I could not see into the nest I was sure there had been a hatch.

I returned home the following evening and went straight to the aviary. On peering in, the female was sat tight and it was obvious she wasn't going to move. Her whole manner had changed though and it was obvious that there were chicks and I decided to leave her alone. As I turned away I detected the smallest of movement from on top of the edge of the nest platform that seemed a bit out of place and not quite right. I remained there watching for about 45 seconds and sure enough, the movement came again. It was then that I saw in amongst the food that had been taken up to the nest that there was a red-tail chick and the movement was a tiny wing. I realised that this chick could not have climbed up there from the bottom of the nest and therefore the female had in all probability thrown it out.

For the next hour I was undecided as to what to do. Should I leave it or should I go in and get it. There must be something wrong with it for it to be thrown out of the nest so what would be the point of going in and upsetting the parents (especially the female who I knew would move in for the kill on me as soon as I entered). Well, after much soul searching I finally made the decision to take the chance and go in. It was still alive, and I couldn't just leave it to die slowly. I also didn't want the parent birds to end up eating it and possibly putting the other chicks at risk.

Well in I went. The female to my surprise only raised herself off the nest and stared intently at me. I had just enough time to see two other chicks in the nest before I grabbed the one on the edge and retreated out again.

The chick was stone cold, but was alive. I also noticed small red marks around the neck area, which may have been consistent with the female picking it up. At this stage it needed warmth and needed it quickly if it was to stand any chance of survival. This is where my wife Lynn came to the rescue and placed the chick near to her heart between those items us men do not have. There it stayed for the next 12 hours and by the morning it appeared to be well on the road to recovery. It managed to take some food and was then placed into a brooder box.

Over the next few days the chick was monitored. The Vet also saw it and nothing obvious was found. It appeared to be perfectly healthy.

On the 8th day I made the decision to return the chick to the nest. I did not want, under any circumstances an imprinted red-tail. I was not sure if the parents would accept it back but I decided to take the chance and see what

Paul Beecroft's daughter, Lucy, with "Leslie", the tough and little fighter male red-tail, who after being abandoned and thrown out of the nest quickly recovered with the Beecroft's loving care and has grown-up to be a beautiful and healthy (non-imprinted!) red-tail. Photo by Paul Beecroft.

would happen. On entering the aviary the female came straight off the nest and launched herself at me. I immediately offered her my gloved hand, which she accepted and locked onto straight away. Whilst she was busy attacking that I slipped the chick back into the nest, disengaged the female and left the aviary. From the peephole I watched her return to the nest platform and then settle back down onto the nest. There were no problems whatsoever and she readily accepted it back again. This bird is now a fully-grown red-tail that turned out to be the one and only Male.

Now the above episode is not the only reason for writing this article. During the time that I monitored the breeding and rearing of the chicks I witnessed an incident or two that to me personally I had not encountered before and I am hoping that some more experienced breeders will be able to comment on this. I have already mentioned that from time to time I would see the male incubating on his own. This was normally only for a very brief time during a break by the female. Also as mentioned I saw them both in the nest at the same time lying side by side.

Following the hatch of the chicks the attitude of the female towards the male changed. A number of times I saw him attempting (or what appeared to me) to get into the nest with the female. She would not have any of this and would push him back out again using her head to do so. She would allow him to stand on the edge of the nest but that was as close as she would allow. After the first week however her attitude changed again and I witnessed a number of times both of

Red-tail chicks, about 7–8 weeks old in nest. Photo by Paul Beecroft

the parents brooding the chicks together. However, the main surprise came when I saw the female tearing at the food, passing it to the male who in turn, then fed the chicks. I saw this occur a number of times and to me it was a surprise. In the main however the female did most of the feeding as would be expected.

They have now bred regularly every single year and I have since 1998 been able to monitor them on a CCTV camera. They are quite incredible to watch. The male is constantly preening the female whilst she sits on the nest and when he wants his turn he will gently push her off with his head. He is not able to cover three eggs properly and spends sometime trying to sort them out but one egg normally ends up not being covered in total. She will normally only allow him 10 minutes or so before she then returns and pushes him off just as gently. Hatching of eggs is not on a 48 hour basis. She does not fully incubate until after the second egg is laid and only sits at night leaving them uncovered during the day.

Normally two chicks will hatch within 24 hours and some 36 days after the first egg is laid although 39 days has occurred before the first chick has arrived. The feeding of the chicks remains with both birds taking turns to feed and one or the other will normally tear off small pieces, give it to the other, who will then feed it to the chicks. The male will also take turns in covering them and I will still see them on a regular basis sitting side by side in the nest.

The Red-tailed Hawk

Buteo jamaicensis borealis. Adult red-tailed hawk, Harrison, MO, USA (November). Photograph by Brian K. Wheeler

2. Subspecies of the Buteo jamaicensis

Thus far, 14 different subspecies of Buteo jamaicensis have been recognized, not counting of course variations in colour—or "colour morphs"—*within* each of the subspecies, or cases of albinism.

For this book, it has been my privilege to have at my disposal the expertise and wisdom of Juan Manuel Iglesias, a master falconer from Spain residing in Puerto Rico, who has assisted me in classifying the various subspecies found on the island. The last of these, has been found in Puerto Rico, where Iglesias operates a refuge for birds of prey, and has studied the variants of the subspecies found on the island for over more than twenty years. Apart from classifying the varieties of red-tail, he has discovered a completely new subspecies of the red-tail that has not been officially recognised and with unique characteristics that Iglesias will discuss in some detail in a forthcoming book. This hawk that he proposes as a subspecies, the Buteo jamaicensis portoricensis, is native to the island of Puerto Rico, and is the smallest subspecies of the red-tailed hawk.

The fifteen subspecies of the red-tailed hawk, with the names of their discoverers in parentheses, are shown in the inset.

There are authors, however, such as Todd (1950), Dickerman and Parkes who have considered the existence of other possible subspecies, for example, the *Buteo jamaicensis abieticola* and the *Buteo jamaicensis suttoni*, although neither of these subspecies[1] have been officially recognised, since so few examples of each have been identified, and it is thus uncertain as to whether each constitutes a subspecies rather than a minor variation of Buteo jamaicensis.

Buteo jamaicensis abieticola received its name from Todd (1950), and were first discovered in a vast zone of northern Canada extending from Alberta in the west to Nova Scotia in the east. The birds of this subspecies

> **The 15 subspecies of the red-tailed hawk:**
>
> B.j. alascensis (Grinnell)
> B.j. harlani (Audubon)
> B.j. calurus (Cassin)
> B.j. kriderii (Hoopes)
> B.j. borealis (Gmelin)
> B.j. umbrinus (Bangs)
> B.j. fuertesi (Sutton & Van Tyne)
> B.j. fumosus (Nelson)
> B.j. socorroensi (Nelson)
> B.j. hadropus (Storer)
> B.j. kemsiesi (Oberholser)
> B.j. costaricensis (Ridgway)
> B.j. jamaicensis (Gmelin)
> B.j. solitudinis (Barbour)
> **B.j. portoricensis (J.M. Iglesias)**

---

[1] According to C.R. Preston y R.D. Beane in the subspecies section of their work on red-tails: The Red-tailed Hawk, The Birds of North America, No. 52, 1993

are darker and have more markings than other eastern subspecies, such as the Buteo Borealis, and have some typical markings found in western red-tails such as the B.j. calurus.

The second proposed subspecies, B.j. suttoni (Dickerman 1994), has been observed in Baja California, Mexico. These birds are quite a bit smaller than those that have been classified as B.j. calurus or western red-tails. B.j. suttoni appears to be a rare species, and is the subject of controversy among ornithologists, with some contending that it does not constitute a distinct subspecies at all.

In addition to the previously described subspecies and their different colour variations—or colour morphs—there are others that have been called "intergrades". These latter represent crosses between different subspecies and are also called natural hybrids. They occur frequently among different subspecies of Buteo jamaicensis and even at times between specimens of Buteo jamaicensis and other hawks. The existence of so many different cross-breeds makes classification a challenge, although it does result in very interesting combinations, as we shall see later on.

Generally speaking, the colour of their feathers depends on the subspecies (with individual variations). Colours may range from dark Brown (almost black) to a very pure white (see next chapter on subspecies). Albinism tends to be observed more frequently in red-tails than in other species of hawk. According to the most recent guides of North American birds of prey, published by Brian Wheeler, albinism is defined as the presence of a much smaller than normal amount—or the complete absence—of pigment in the eyes, skin or feathers. There are four kinds of albinism:[2]

1. **Total albinism** – Where a complete absence of normal pigmentation in the eyes and skin is present, with both areas being pink. Here the feathers are pure white. This is the rarest form and rarely seen in raptors.

2. **Incomplete albinism** – Normal pigmentation is completely absent in eyes, skin or feathers but not in all three areas. Plumage is often pure white or nearly so, but in most cases eye and skin colour are normal. However talons are often pink coloured.

3. **Imperfect albinism** – Normal pigmentation is only partially reduced in eyes, skin or feathers but not totally in any of the three areas. Also known as "dilute plumage". Eye pigmentation is either normal or bluish. Skin areas are normal or slightly paler than normal in colour. Most or all of the plumage is a tan colour; any typical markings that have a rufous or tawny colour are altered to a pale rusty colour in this plumage.

4. **Partial albinism** – Normal pigmentation is completely lacking on portions of the body. Eye and skin areas are usually of normal colouration. Plumage is often a patchwork of normal and white feathers, or parts

---

[2] Information obtained from the works of Brian Wheeler "Raptors of Eastern North America" and "Raptors of Western North America", published with the author's permission.

2. Subspecies of the Buteo jamaicensis

of several feathers may be white and other parts normal. Most common type of albinism.

Albino birds are occasionally seen throughout the United States, although they tend to be observed more frequently in the east than in the west. They are most rare in California and in the southern states, where darker subspecies are most frequently observed.

An interesting characteristic of these albino birds is that they for the most part do not migrate southward. They stay within the territory where they hatched, and their self-defined zones tend to be very small in comparison to the zones of other birds. No red-tail couple has yet been observed in which one of the pair is completely white.

According to the article of Mike Faueux on albino red-tails, it appears that albino specimens tend to be driven out of the territories of their non-albino brethren. However, partially albino specimens have been found to mate with non-albino red-tails. Mike mentions in his article that the tails of partially albino birds can sometimes turn red in colour, and then grow white again after the next moulting.

Not much is really known about albino red-tails, although a number of articles have been written and a study focusing specifically on albinos raised in captivity was begun in 1997, although the results of this study are as yet unknown. The object of the study was to raise a number of albino birds to be used in falconry.

At present, the little information we have about albino red-tails based on the studies that have been carried out reveals that they have a greater tendency to suffer from diseases, and they have very limited defences. Specimens with light eyes also have much lower visual acuity than those with dark eyes. As regards size, they are not distinct from other red-tails, according to an article written by Charlotte and Roy Lukes. In fact, all seven sizes of ankle ring may be utilized to identify albino red-tails.

All of the foregoing leads to the conclusion that these totally albino specimens are difficult to find not only because cases of total albinism are rare, but also because these specimens have lower rates of survival and breeding because of their genetic weakness and their rejection by non-albino red-tailed hawks. The latter phenomenon results in their genes not being transmitted. I think that it is a shame that more studies are not carried out on these fascinating and beautiful specimens as well as the different red-tail subspecies.

Returning to the topic of subspecies, these may be grouped into two broad categories according to geographical distribution: Western (including the subspecies calurus and fuertesi) and Eastern (including umbrinus, borealis, and kriderii), with a number of common characteristics observed within each of these two sub groupings.

As is the case with other birds of prey, red-tailed hawks tend to decrease in size as one moves from southward from the northern arctic reaches of the Canadian tundra, across the border into the northern US, across the Mason-

The Red-tailed Hawk

Dixon Line, and into the Deep South. Northern birds are therefore larger than southern birds. Another characteristic that has been observed is that specimens from the south-western US usually have incomplete dark stripes in their tails that are narrower than the tail's subterminal band. This characteristic is peculiar to the calurus, but has also been observed in other southern red-tails such as the portoricensis and the jamaicensis.

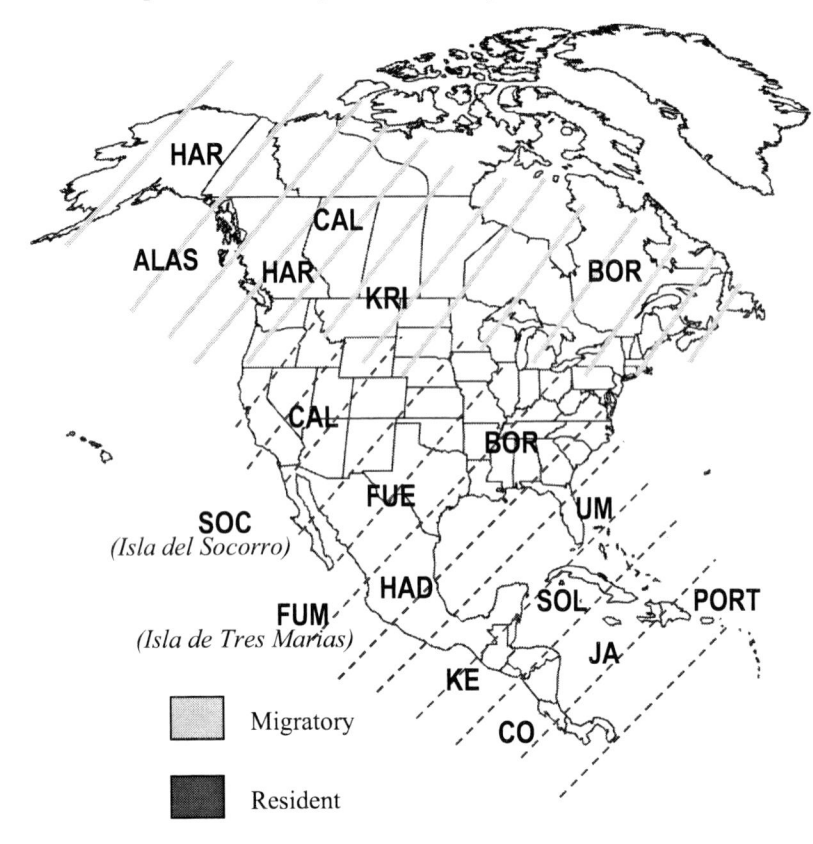

Map of geographic distribution of red-tails in North America, adapted from Johnsgard and Preston.

ALAS = Alascensis	FUE = Fuertesi	JA = Jamaicensis	SOL = Solitudinus
BOR = Borealis,	FUM = Fumosus	KE = Kemsies	SOC = Socorroensi
CAL = Calurus	HAD = Hadropus	KRI = Kriderii	UM = Umbrinus
CO = Costaricensis	HAR = Harlani	PORT = Portoricensis	

West
- Darker, more rufous colouring
- Shorter, rounder wings
- Short, fan-like tail
- More similar to other buteos (common buzzard)

East
- Lighter colouring
- Broader and longer wings
- Longer tail
- More similar to accipiters

2. Subspecies of the Buteo jamaicensis

As for the characteristics shared among all red-tailed hawks, there are three[3] common elements shared by all red-tailed hawks (except for harlanis).

- The reddish tail of adult members of the species.
- The two tones in the superior section of the wings in juveniles.
- The darker coloured markings underneath the wings, at the level of the shoulders on the outer part of the wings, in red-tailed hawks with lighter plumage.

The plumage of males and females is similar and, contrary to the case of other birds of prey, it is not possible to determine sex based on plumage alone, although it is often true that some females are darker or have more defined marks, but this is not always the case, as I have seen for myself with the red-tails that I have owned.

Weight is also not a reliable indicator because, even though females are generally larger than males, as is the case with the majority of birds of prey, there are many exceptions to this rule. There have even been instances of males who have—based upon size—erroneously been classified as females, only to later surprise their classifiers by laying eggs. There is also a great variation in size, weight and colour not only among subspecies, but even within a single subspecies. There are, in addition, great differences in plumage among subspecies and natural hybrids—differences that are also observed among juvenile birds. Some juvenile birds of the species are nearly identical to adult specimens, the chief difference being that the former lack the reddish tail. In addition, the following constitute essential and commonly observed differences between juvenile and adult red-tails:

- Eye colour, as we have seen, changes with age, and juveniles tend to have lighter-coloured eyes that are of a greyish yellow hue. Red-tails' eyes tend to darken with age, eventually acquiring a maroon or chestnut colour.
- The cere around the beak of juveniles tends to be dull yellow or green in colour, and generally matches the feet in colour.
- In many young red-tails the abdominal belly band is more defined than it later becomes in adults, often being darker and more pronounced in the young.
- Juveniles have longer feathers, and therefore their tails of juveniles are much larger than those of adults, and adults often have wider wings.

As mentioned previously, the Buteo jamaicensis has been subdivided for purposes of classification into 14 officially recognized subspecies and one newly discovered apparent new subspecies, with each of these subspecies displaying one or more important differences from birds belonging to the 14 other subspecies. We will now turn to a presentation of these differences.

NOTE: *With regard to these variations, it is interesting to note that, in*

---

3 According to the Peterson Field Guide: Hawks of North America, by William S. Clark and Brian K. Wheeler, 2001.

contrast to the red-tails observed on the US mainland, those in Puerto Rico tend to be ornithophagous, possibly because there is a higher diversity of rodents in the continental US than there are in the Caribbean isles. Whatever the reason for this difference in eating patterns, it is yet another instance of the adaptability of these birds. In fact, biologists Noel and Helen Snyder conducted research on the cotorra of Puerto Rico (*Amazona vittata vittata*) in 1975 for the El Yunque National Forest. During their study, they observed a nest of *Buteo jamaicensis* birds, and the prey they had brought back there. In addition to centipedes, rats, reptiles and doves, they discovered that these island red-tails often captured the Puerto Rican cotorras (both chicks and adults). As we have already seen, and as we shall see yet again throughout this chapter, *Buteo jamaicensis* adapts very easily to different territories and birds of prey. This allows the bird to thrive in widely varying climates. However, it remains true that red-tailed hawks primarily hunt mammals.

Subspecies in the US and Canada

Buteo jamaicensis alascensis (Grinnell)

This subspecies of red-tail is from the south-eastern part of the state of Alaska (US) towards Vancouver (Canada). Its name in fact means "Alaskan red-tailed hawk." These birds are typically migratory, flying southward into the south-western areas of the Canadian province of British Columbia.

They may crossbreed with calurus red-tails that reside in the central areas of British Columbia, producing "intergrades".

Adults of the species bear a strong resemblance to the darker adult specimens of the calurus subspecies, also native to western North America, but have a reddish hue that is more intense on the wings, both on the inferior surface and on the wing coverts beneath the wing. The adult birds on Queen Charlotte Island have a characteristic arrow-shaped marking on the chest. They almost always have a dark throat or bib, and their belly-band is usually very dark and has markings, with their sides, abdomens, and feathers of their thighs having brown streaks. The bands or lines in the tails are very wide, although it is possible that only certain individual birds have a dark sub-terminal line.

The young of this subspecies are very similar to the young of the calurus subspecies, but with a head and upper body that is dark brown—in fact, almost black at times. In addition, young alascensis birds have throats or bibs with lines that are more clearly defined, as well as dark bands or lines in the tail that are wider than those of those of calurus.

Buteo jamaicensis harlani (Audubon)

Buteo jamaicensis harlani is also known as Harlan's hawk, having been thus named by the great naturalist J.J. Audubon after his friend Richard Harlan. This bird, which is also known as a "black warrior", has been the subject of controversy as to whether it constitutes a different species (Mindell, 1983), or whether it is merely a melanistic

2. Subspecies of the Buteo jamaicensis

(i.e., darker coloured) variant of calurus[4] (Palmer, 1988). It was not recognised by the American Ornithologists Union (AOU) as a true subspecies of red-tail until 1973[5]. Even after that recognition, there continues to be division among ornithologists. Two reasons why some suspect that it is a variant of calurus are, first, that the two types of birds are very similar in appearance during their early years and, second, that they frequently interbreed with one another, producing intergrades.

This subspecies of Buteo jamaicensis has its breeding grounds in Alaska, from Norton Sound (in the far northwest of Alaska) down to the southern part of the state, from the southeast of the Yukon Territory (the province of Canada bordering Alaska on the east) to the Northern part of British Columbia, interbreeding in various zones with calurus, kriderii, borealis and probably also with alascensis. It migrates south into Kansas and Missouri and sometimes as far south as the Gulf of Mexico. Specimens have been observed in Oklahoma, Texas, and Louisiana.

This red-tail is in its standard and most widely known version, the darkest of all the red-tailed hawks, not taking into account melanistic forms of the harlani or of other subspecies. In its standard version, the colour of the harlani ranges from dark chocolate to an almost black shade of brown. This subspecies is the only one that does not have a tail that is actually red (not counting albino birds), the tail instead being a grey-brown colour with a little bit of red, and with blank specks or blotches, with a fair amount of individual variation. Their size is slightly smaller than that of the western calurus, and their necks are thinner, and their hands more delicate. The younger birds of this subspecies have a tail longer than that of calurus, and resemble the Buteo lagopus. Adults do not have the typical abdominal stripe of typical adult red tails, and they usually have white specks on their chest.

Similar to the other subspecies of the red-tail, the harlani displays a great deal of variation in colour, and it is impossible to determine their sex on the basis of their plumage and weight alone, and there is a great deal of variation in both these variables among both adults and juveniles. Only after the first moult will they obtain a darker or lighter adult plumage (depending on their colour phase).

There has never been an instance of albinism reported in this subspecies and, although there are instances of lighter coloured individuals among this species (and such specimens have been at times confused with the kriderii subspecies, the vast majority of these birds are rather dark. In cases of crossbreeding with other red-tails, the offspring have been observed to have characteristics of both subspecies. Some individual specimens that have resulted from the crossbreeding of harlani with the Buteo lagopus[6] have been reported.

4–5 Paul A. Johnsgard, (1990), "Hawks, Eagles & Falcons of North America".

6 In the Peterson Field Guide to Hawks of North America by William S. Clark and Brian K. Wheeler, second edition, 2001, there is a photograph of what could possibly be a natural hybrid of a harlani with a rough-legged (Buteo lagopus).

The Red-tailed Hawk

As for colouration, there are five[7] different recognised shades of colour among harlani hawks:

- **Light phase** – (rarely observed)

    Harlani having light colouration resemble kriderii hawks, although they don't have the same shade of brown. Some individual specimens appear silvery, lacking any brown colour at all, with colouration ranging from white to dark grey/black. The heads of such birds are very speckled with dark brown and white spots, and the face and throat are white. The wing coverts of the back are very dark brown, with white specks that form a "V" shape on the back. The proximal part of these birds' bodies are white, with lightly marked belly-band (with some specimens displaying a more pronounced marking).

    These differences between light-coloured harlani birds and kriderii are particularly noticeable upon observing these two different species of red-tail in flight. The colour of the tails of light-coloured birds varies considerably from white to dark grey, normally with a darker subterminal streak. There may also be read highlights near the tip of the tail, and dark speckling and/or streaking. Some tails have many incomplete streaks, and a wider subterminal band.

    Light-coloured juvenile harlani are very similar in appearance to adults, except for having a more clearly defined ventral area (belly band); the head (which is not white) is usually dark as well (but without the brown tones of the kriderii, instead it appears as that of a grey red-tail) and the tail is brown and barred just like all juvenile red-tails but subterminal band ends in a sharp point similar to a spike.

- **Moderately light birds** – (less common)

    These birds are similar in appearance to the light-coloured harlani, although they exhibit dark streaks on their white chests, a darker chest, dark speckling on their sides, and some dark markings under their wings. Dark markings under the shoulders are also noticeable, and in this area of the body, moderately light juveniles are similar in appearance to the previously described light-coloured birds, although they have a larger number of markings.

- **Intermediate birds** – (very common, the average typical harlani)

    The head is dark, but the eyebrow and throat are white in colour. The chest is either striped or speckled in white and dark colours. The abdomen has white markings, as do the areas underneath the wings. The feathers on the upper part of the tail may be reddish in colour. The red-tail has not distinctive markings underneath its shoulders.

- **Moderately dark** – (very common)

    Birds of this colouration present with very little white speckling, and what there is of such speckling is limited to the chest and, sometimes, the throat. In some specimens, there is a dark chest

---

[7] According to Brian Wheeler, in his The Wheeler Guide. Raptors of Western North America. Princeton University Press 2003.

2. Subspecies of the Buteo jamaicensis

without speckling, but with some white speckling in the abdominal region.

- **Dark** – (less common)
Birds of this colouration do not display white speckling but instead generally have a dark brown/black colouration.

Typical dark juvenile harlani specimens strongly resemble the juveniles of other red-tail subspecies, especially the calurus, with the exception of the colour of the head and upper body being generally much darker. The bird thus has the appearance of being a "darker sort of red-tail hawk" (when they hatch, they are already much darker than typical red-tails, and display a colouration that is completely different from that of other red-tails, as can be observed in the accompanying photos. There is no ochre colouring on the head, neck or chest. They eyes are generally a pale shade of grey, sometimes with a yellowish tint: In this and other ways, they share the characteristics of juveniles of other red-tail subspecies that have been previously discussed here.

It is interesting to note that the tails of harlani juveniles resemble the tails of other red-tail juveniles, except for the final barring or subterminal band, on the tail, which ends in a sharp point as if it were a spike. This exclusive feature can be observed in some of the secondary feathers and seems to be present in harlani juveniles or natural hybrids of harlani with other red-tails.

The dark-shaded harlani birds have often been confused with other hawks such as, for example, the ferruginous hawk (Buteo regalis) since the darker coloured ferruginous adults may closely resemble the harlani, having white-speckled chests, as well as other characteristics, in common with them. They have also been confused with the rough-legged hawk, and the two types of hawks often do in fact crossbreed. The rough-legged hawk (Buteo lagopus) is very similar in appearance to adult dark harlani birds, with the exception of the fine details of the tail.

Buteo jamaicensis calurus (Cassin)

This is the most well known subspecies of red-tailed hawk in Spain and other Mediterranean countries for falconry. It is also present in the United Kingdom and other European countries. The relatively scarce examples of red-tailed hawks in the UK are originally native to the western US, and are known in their original home territory as "western red-tailed hawk." They take their name from the Greek word "calos", meaning "beautiful", and "oura", meaning tail. The name of the bird thus has the literal meaning of "beautiful tail"[8]. This reddish-coloured subspecies reproduces in a zone extending from Central Alaska and the northern Yukon to (where it appears to crossbreed with the harlani) down the west coast of the US and into Baja California, Arizona, New Mexico and Texas (see distribution map). They usually spend the winter migrating south (Arkansas, California, Louisiana, Missouri and Texas), occasionally to Mexico and Panama, although like other red-tail

---

[8] According to Paul A. Johnsgard's "Hawks, Eagles and Falcons of North America", Smithsonian Institution Press, 1990.

subspecies, southern populations are not usually migratory.

The calurus is the subspecies most commonly found in the western US, just as the borealis is the most common subspecies of the eastern US. The calurus is of an average size compared to other red-tails, with adult males usually weighing 800–900g (28.2–31.7 oz) and the females weighing an average of 1,500 grams or 52.9 ounces (from 900 to 1,500g / 31.7–52.9 oz).

This subspecies is reddish brown in colour, having a dark-coloured throat and upper body. Its sides, abdomen and the feathers of its thighs may have rust-coloured barring. There are three basic varieties of colouration displayed by this subspecies, with many variations with each of these due to the presence of intergrades: there is a light-colouring, a reddish or "rufous" colouring (in which the typical markings of the red-tail are still visible, and are of a chocolate or cinnamon colour) and a third, dark, colouring (which is most typical of the western subspecies) in which the markings are rather indistinct. The tail of the calurus is somewhat shorter than that of the eastern subspecies, and is shaped like a fan. The tail of adult calurus birds is usually reddish in colour, as is typical of red-tailed hawks in general, and there are other narrow stripe markings at the end of the tail.

It is interesting to note that newly hatched chicks and young birds of the calurus subspecies closely resemble their harlani counterparts, and that there are many intergrades among these two subspecies, as well as with other subspecies. For this reason, they are very difficult to classify.

This subspecies is also characterized—as are all western subspecies—by its wider wings and shorter tail in comparison to eastern subspecies. As for its appearance, it tends to be broader and rounder than eastern subspecies, and is therefore more similar in appearance to a buzzard. It has hands there are smaller and more delicate, and that are not as powerful, as those of eastern red-tails. Still, the calurus is perfectly capable of hunting fairly large female hares, as long as it is properly trained. The calurus is used for this purpose in the US, UK, Italy, and other countries.

In Mediterranean countries especially, the calurus is usually not used in falconry, but is instead raised in captivity in very small numbers, displayed in zoos, or shown in falconry exhibitions, such as the Madrid Safari, where it can be seen flying long distances from a watch tower towards a dragged lure (rabbit lure).

As for hunting, the calurus flies rather more slowly than other subspecies like the Buteo jamaicensis kriderii or borealis, and although it can do a good job of hunting rabbits and hares, it may have some difficulty with the latter—especially if they are larger than average. Its hands are not as powerful as those of the fuertesi or kriderii subspecies, and it is not very bulky. In general, it cannot be used to hunt squirrels. In Spain, the calurus is rarely used to hunt, but in England, where they comprise the majority of red-tails, they are often used to hunt rabbits and hares.

There is evidence of intergrades or natural hybrids with all subspecies of red-tailed hawks to which the calurus lives in close proximity, especially the harlani and borealis. In these natural hybrids, there is often a particular mixture of physical characteristics reflecting the mixed lineage. In harlani-calurus hybrids, for example, the head and chest are generally similar to the harlani. On the other hand, the underside of the shoulders will usually have the reddish colour characteristics of calurus. In borealis-calurus hybrids, the feathers of the thighs are often barred, the throat is often some shade of red, and the ventral area is more pronounced.

Buteo jamaicensis kriderii (Hoopes)

This subspecies is called Krider's Hawk in English, and receives its name from John Krider, who collected examples of subspecies in Iowa (US). Krider's Hawks are prevalent in American and Canadian plains. They were initially classified as an albino form calurus or borealis, and were not recognized by writers such as Palmer (1988), since they display great variation in their colouration. Generally, they are the lightest coloured subspecies of red-tail, and some specimens appear to be completely albino, since they are entirely white in colour. It is one of the largest and heaviest of the Buteo jamaicensis subspecies—and, in my opinion, one of the most beautiful.

The kriderii red-tail is very similar to other red-tail subspecies, although it has a very faded colour—with some specimens almost entirely white and others appearing to be muted versions of the typical eastern red-tail. It is thought that they often produce intergrades with other subspecies sharing the same territorial subspecies, especially the borealis. This subspecies of red-tail has frequently been confused with the light-coloured harlani, with light coloured fuertesi, and with the ferruginous hawk (Buteo regalis).

This is the red-tail of the great plains of Canada, Wyoming and Nebraska where it often crossbreeds with the Buteo jamaicensis borealis, both of these subspecies having "eastern traits"—i.e., shorter, rounder, and more slender than other red-tails. These two subspecies tend to produce hybrids that are large in size and strikingly beautiful in appearance. The kriderii red-tail also often produces hybrids with calurus within the western and northern areas of its residence on the American continent.

This subspecies is generally migratory, travelling south during the winter to the Gulf coast, mainly Florida but also to Georgia. Some individuals have also been seen down in Mexico.

In terms of physical appearance, the kriderii red-tail is identical to the borealis, but it is much whiter in colour. Seen from the ground, it appears almost entirely white, except for some brown markings on the inside of its shoulders. The tail of the Krider hawk is variable in colour, ranging from a very bright white to a pinkish white to bright red, with a fine streak at the tail's end. As is the case with the subspecies of the east, the coverts and

The Red-tailed Hawk

the back are brown with white specks. The abdomen and the sides are white, with no markings or specks of any kind. The Krider hawk is similar to any eastern red-tail in having white splotches on the upper part of its body, on its head, and on its wings. It can be confused with the ferruginous hawk (Buteo regalis) at times, although the two obviously differ from one another in shape and size.

Juveniles of this subspecies typically have a white head with no markings except for those on the base of the neck. The back is very similar to that of the borealis, although it is of a shade of brown that is much lighter, and displays a great deal of white speckling. The juveniles of this subspecies always have some barring on their sides and abdomens. Their tails are also light brown in colour, with barring similar to that observed among the juveniles of other subspecies. Part of the tail may also be white.

This subspecies generally has very large feet and thus, like the fuertesi, is particularly well suited for hunting not only hares and rabbits, but squirrels as well. The female especially possesses the natural endowments for hunting such animals, but larger males with large feet are used for the same purposes, and have the advantage of being more agile than females.

The Krider hawk has at times been confused with the borealis subspecies, and it is easy to understand the reasons for this confusion, given that the two variants share many of the same living environments and frequently interbreed with one another. Hybrids of these two subspecies (classified by some as examples of Buteo jamaicensis kriderii/borealis) may have heads that are darker than those of pure kriderii birds (remembering that pure kriderii red-tails have heads that are nearly completely white in colour), while retaining tails that are typical of their subspecies (i.e., white or pink in colour). The eyebrows of the kriderii may be of a pronounced white colour in specimens that have darker heads.

Buteo jamaicensis borealis (Gmelin)

The Latin name of this subspecies literally means "northern red-tailed hawk", and it is native to the north-eastern US. It is also found in the plain states that are home to the kriderii. The members of this subspecies that breed in northern climates are generally migratory, while those that breed in the south do not typically migrate. There are, however, examples of northern birds that do not migrate southward for the winter. As with the kriderii, those that do migrate south make their winter homes in Florida and neighbouring south-eastern states.

This subspecies is also known as the "eastern red-tail, because it is one of the most common subspecies in the eastern United States, and is one of the subspecies most commonly used in falconry. This is the subspecies that most closely resembles an accipiter—and, particularly, the goshawk. The borealis is usually lighter than other red-tail subspecies, with males weighing an average of about 1kg (35 oz) and females an average of 1.2kg (42.3 oz). However, there are individual specimens of this subspecies that are much larger than this ; my male

2. Subspecies of the Buteo jamaicensis

red-tail for example was close to 1600g, being a predominantly eastern specimen.

Physically, these red-tails are thinner than their western counterparts, with shorter and rounder wings and a longer tail. It thus doesn't seem entirely appropriate to include borealis in the group of "broad-winged hawks" that includes the common buzzard, the ferruginous hawk, the Harris hawk, and the red-tail hawk. In fact, borealis generally bears little resemblance to a "typical" hawk—unlike the western subspecies. Moreover, its manner of flying is similar to that of the goshawk, not only in its form and speed, but in the way it hunts its prey. Normally, red-tails sit atop telephone poles near highways and sometimes hunt from those positions. Borealis seems to be the only exception to this rule. It displays a temperament that is also similar to that of the goshawk, although they are somewhat easier to handle than the latter.

The heads of the borealis red tails are generally brown in colour, with streaks of a darker brown and a white throat that is marked off by a "collar" or line. The wing coverts of the back and the upper surface of their wings are dark brown in colour with white specks that clearly form a "V" shape on the back, with white specks on the upper wing coverts. On the lower part of the body, there is an incomplete abdominal boundary. This species of red-tail does not generally have a dark-coloured (i.e., melanistic) variant, as is found among western subspecies such as calurus or harlani.

The plumage of juvenile borealis red-tails is very similar to that of the adults, although the brown of the head is lighter and the eyebrows are white, with the throat often being bounded by what looks like a narrow "collar. The belly band of the borealis is generally light in colour but with dark markings, much more so than is the case with adults. The feathers of the thighs may or may not have barring. There is a great deal of white streaking on the back, and which forms a "V" shape which can clearly be seen as well as the "U-shape" made up by white speckling just before the tail. Juvenile specimens have tails that greyish brown in colour with horizontal barring of a darker colour, terminating in a wider and darker band, and with the very tip of the tail that is white.

As is the case with other subspecies, the borealis is very adaptable, and has even been found in Puerto Rico. The Puerto Rican borealis has been termed the "long eastern red-tail" by J.M. Iglesias. This variant is slightly smaller than the typical borealis of the east, with males weighing 750g (26.4 oz) and females about 1kg (35 oz). It has large, well-developed feet and is an excellent hunter.

Puerto Rican variant of the borealis subspecies

The Puerto Rican variant of this subspecies is thinner more elongated in form than other red-tail subspecies. As explained, males weight about 750g (26.4 oz) and females about 1kg (35 oz). This bird is more similar to an accipiter than a buteo: It has a small head, the gaze of an accipiter, wings that are shorter than a tail which is typically longer and wider than other birds of the same subspecies on the mainland.

The tail of these specimens is larger and wider than that of other subspecies. In juveniles, there is a wide black band at the end of the tail, along with a striking off-white colouring in the rest of the tail. The throat is white. The back of the bird has white markings, and the ends of the wing coverts are white. The tarsi are long, the feet are large and well developed, and the birds of this subspecies are good hunters.

The Puerto Rican borealis has a quiet and submissive temperament, and though can be a bit of a screamer, is a very self-confident bird. When it does decide to act, it does not hesitate, although it never displays a great deal of emotion.

It flies very fast and erratically: When it pursues its prey and the latter accelerates, it will generally abandon it within the first few metres (6–8ft). It shows a preference for either small forests where there are vast scrub-lands, where it takes great advantage of dense growth to lie in wait for its prey. It also frequently uses the technique of flying low in search of prey that are also hunted by accipiters, using this method to surprise its victims. Using this technique in the same way as the goshawk, it can fly up to 600 metres (about 656yrd) without getting tired.

As for their habitat, they very much like areas with the type of growth that favours their particular style of flying. During the years that Juan Manuel has worked with them, he has never seen them sit atop telephone or electric poles, even when such poles were present in their environment: Instead, they always prefer trees.

Buteo jamaicensis umbrinus (Bangs)

This red-tail, which is also called the "Florida Red-tailed Hawk, is the variant of red-tail that is most prevalent in the state of Florida (the borealis being the most common red-tail in all other areas along the east cost of the US). It can also be found on the larger islands of the Bahamas.

This subspecies does not migrate, and shares its territory for part of the years with northern subspecies like the kriderii and the borealis that do migrate southward.

Adults of this subspecies are very similar in appearance to the calurus, although they are darker (in fact, the Latin name of this red-tail literally means "dark Jamaican hawk". Its head and the base of its neck are dark brown in colour (darker than the borealis, similar to the calurus) and it is usually not possible to see a clear "V" on the back—something that can be done with lighter-coloured species. The belly-band as well as the sides is very dark in tone (in other subspecies, the colouration is usually not so dark). The chest has no markings. The middle of the abdomen is white and does not have markings. The feathers of the thighs are white and may or may not have barring. The wings, however, of its upper body do not usually display white specks (in contrast to borealis and calurus). The markings underneath the shoulders are very pronounced and are very dark—almost black—in colour. As is the case with other southern red-tails, the tail has an intense reddish tone with a dark and wide subterminal stripe in addition to other small stripes, narrower and incomplete, similar to the tail of the calurus.

The juveniles of this subspecies are identical to the juveniles of borealis. Thus, it is not possible to differentiate the young of these two subspecies until their plumage changes.

Puerto Rican variant of the umbrinus

There is a variant of the umbrinus in Puerto Rico that has been classified by Juan Manuel Iglesias, and which he calls "compact". The males of this variant weigh an average of 900g (32 oz) and the females weigh and average of 1,400g (49 oz).

Compared with other red-tail variants present on the island, they represent an intermediate type between the "strong" red-tails (i.e., the Puerto Rican variant of fuertesi) and the "long" red-tails (the island variant of borealis, previously discussed). It has greater volume than the former, and has a rounder form and higher pectoral mass than the latter.

The primary feathers of this subspecies are shorter than the end of its tail. Its head is small, its neck long, and it has a convex chest and a generally rounded profile. The tarsi are of medium size and the feet are large, but not nearly as strong as those of the fuertesi. It displays a great deal of strength. Its gaze is not also as piercing as that of the fuertesi, and is instead rather gentle. In appearance, it closely resembles the eagle.

These red tails are alert and are not overly aggressive; however they do often appear uneasy. They are also not as noisy and conspicuous as the borealis.

Its flight does not show a high degree of poise and refinement, but it has a great deal of spirit, and it will fearlessly go after any prey whatsoever. It seems that his subspecies is congenitally anxious. When the falconer handles it, it will often try to cling to his glove, thus, it will often end up hanging on the falconer's glove when it tries to take off. This also happens when it is perched on a branch in a tree—it can sometimes appear that the buteo is trying to break off the branch. Because of this defect, these birds pose a management problem for the falconer, as the distracted falconer might end up with the bird on its face—or hanging from some other part of his body.

The flight of the Puerto Rican "compact" variant of the umbrinus is not as strong and firm as that of the fuertesi, and it tends to pursue its prey in short, rapid spurts. I have not yet observed one of these birds in a full sprint. If the prey is strong and fast, it may choose not to pursue it, limiting its flight to rather short distances.

The compact umbrinus prefers areas with lots of trees, including mountain forests. However, as with the case of other subspecies of red-tail, it also likes to perch on wooden or metal poles.

Buteo jamaicensis fuertesi (Sutton & Van Tyne)

The Buteo jamaicensis fuertesi is one of the red-tail subspecies that is most frequently employed in falconry. It is named in honour of Louis Agassiz Fuertes, an American ornithological painter. The fuertesi is one of the larger subspecies of the red-tailed hawk and, like the kriderii, has very powerful feet. Like the kriderii, it is ideal for hunting rabbits and large hares, as well as squirrels.

The Red-tailed Hawk

Individuals of this subspecies are typically found in Texas and Arizona, but they have also been observed in Mexico and Puerto Rico. The Puerto Rican variant, named "Roka" by J.M. Iglesias and of a much darker colouration than the typical fuertesi of Texas and Mexico, is also slightly smaller in size, with males averaging 850g (30 oz) and females 1,150g (41 oz). The fuertesi, like other southern subspecies, is generally non-migratory.

This subspecies may be thought of as a version of an adult calurus with lighter coloured wings. It is also similar to the borealis in some respects, although it has longer wings and generally lacks a belly stripe, and normally has a dark throat or bib (especially in areas where it shares territorial space with western red-tails). It may have diffuse barring that is rust-coloured on the sides, and it may also lack the subterminal marking on the reddish tail. The head is usually completely brown. Seen from the ground, it appears to be very white, similar in this respect to the eastern subspecies, including the kriderii, but somewhat darker, with a reddish tail and brown wing coverts. This subspecies of Buteo jamaicensis is, in some of its tonal variations, reminiscent of the booted eagle in appearance.

Juveniles of the species closely resemble those of the borealis, but with much longer wings. In addition, the juveniles of borealis present with a darker and more marked abdominal zone, and with a throat that is either dark, or that has dark markings.

Puerto Rican variant of the fuertesi

The variety of the fuertesi subspecies in Puerto Rico—given the name "Roka" by Iglesias on account of its particularly strong physical constitution, and whose photos appear on pages 121 and 122, is slightly smaller in size to the fuertesi specimens typically encountered in Texas and Mexico. Males weigh about 850g (30 oz) while females can weigh 1.150g (41 oz) and are darker than typical fuertesi individuals.

From head to tail, the birds have a short, stocky appearance. In adults, the primary feathers nearly reach the end of the tails. They have broad, square shoulders, and the primary and tail feathers appear foreshortened in appearance. Seen from behind, the movement and appearance of this buteo might remind us of the peregrine falcon, with a profound gaze like that of the eagle, although the head is slightly shorter and the feathers on the head (hackles) stand on end. Its thighs are long and thick. Its tarsi are also relatively thick, and it has large strong hands and strong claws. It is aggressive, attentive, and acts spontaneously in pursuing its prey. It is highly spirited and is especially comfortable on land.

The style of flight of the Roka is firm and secure, and suggests the power and elegance of an eagle. On occasion, it will burst into short sprints that resemble those of accipiters, and that usually do not exceed 25 meters (27 yrd). However, they pursue its prey tirelessly, flying low and trying not to lose sight of its quarry, however fast it might be, covering distances of 400 to 1,000 meters (437–1,093 yrd) flying both upwind and downwind. If the prey manages to get away momentarily and then emerges from hiding, it continues its pursuit. Iglesias states that these birds are very energetic, and in this way resemble the large accipiters.

Iglesias has flown individuals of this subspecies in Puerto Rico and told me that a specimen named Roka (who has become famous on falconry forums because of its feats), captured, at a distance of 200 feet (61m) from its original position a pigeon that was originally 100 feet (30m) away, starting from the top of a telephone pole eight metres (26 ft) above ground. Another time, this same specimen captured pigeon after a pursuit of one kilometre (0.6 mi). On yet a third occasion, it pursued a pigeon for 300 metres (328yrd). The pigeon managed to hide, but when it emerged after 15 minutes, it recommenced its pursuit for another 600 metres (656yrd).

The preferences in terms of habitat for this variant are cleared and open areas. It likes to perch atop towering trees or high poles, especially wooden electrical or telephone poles.

The Subspecies of Mexico

In Mexico, in addition to the fuertesi subspecies previously described, there are an additional three subspecies of Buteo jamaicensis that are found only in that country. Very little information is available about these Mexican subtypes, since they don't appear to have generated much interest. I have made numerous attempts to obtain more information about these red-tails from Mexican sources. To my great regret, I have been unable to obtain the detailed information that I was after. These subspecies are now classified as in danger of extinction, and are officially protected by the Mexican authorities. Despite this protected status, there have been reports that these hawks have been the object of illegal trafficking.

NOTE: These hawks are also called "little eagles (aguilillas, in Spanish) and "red-tailed falcons in Mexico (see "Names of Red-tailed Hawks" at the end of this chapter.

Buteo jamaicensis hadropus (Storer)

Originally classified by Robert Storer, its name comes from the Greek term "hadros" which literally means "thick-footed red-tailed hawk" in reference to the large size of its hands.

This Mexican subspecies may found in mountain zones extending from Jalisco southward to Oaxaca. It closely resembles the fuertesi, which is also found in Mexico, although it is somewhat smaller and, unlike the latter, has rust-coloured barring on its sides, abdomen and pendant feathers).

Buteo jamaicensis fumosus (Nelson)

The name of this Mexican subspecies, originally classified by Edward Nelson, comes from the Latin word meaning "smoked", and is a reference to its red colour.

This subspecies, along with the socorroensi, which is described in the following section, is found in the Mexican islands. Both are classified by the Mexican government as being in danger of extinction. The fumosus subspecies is found only on the island of Tres Marías, off the coast of Baja California. It is similar to the calurus that is found in California, but has a rust-coloured belly.

Buteo jamaicensis socorroensi (Nelson)

This subspecies, along with the fumosus, is also found on an island—the Mexican island of Socorro, in the region of Revillagigedo south of Baja California. This island, which has been set aside for use by the Mexican navy, is an inactive volcano, Mount Evermann. The land on this island is composed primarily of dry brush land. Due to its richness of its ecology, it has been set aside by the Mexican government as a protected area. The National Council for the Preservation of Birds as being host to critically important endemic species. However, both the vegetation and the soil of the archipelago have suffered considerable disturbance, and a number of species have been drastically reduced in numbers, with others having become extinct altogether.

On the island of Socorro, the red-tail, or "red-tail sparrow hawk", as it is known locally (Gavilán de Cola Roja), is in threatened with extinction. I think that this is the case mainly because of the scarcity of food on the island, since this bird, in addition to feeding on mammals (that have been few and far between on the island until the recent introduction of rabbits) also feeds on birds, particularly pigeons. The pigeon native to Socorro, the Zenaida graysoni, is thought to be extinct, the last specimen having been observed in 1931. As the red-tail is so adaptable, its diet now consists primarily of rabbits, because various plans have been implemented to eliminate animal species that have caused damage to the environment, such as rabbits, dogs, cats, pigs, terrestrial crabs, lizards, and small invertebrates.

As for its plumage, it is very similar to that of the calurus of California, but darker. It also has larger and more powerful hands than the calurus (similar to the fumosus).

NOTE: All previously mentioned Mexican red-tail subspecies are resident (non-migratory).

Guatemala, Nicaragua, Panama and Costa Rica

Buteo jamaicensis kemsiesi (Oberholser)

Named by the American naturalist Harry C. Oberholser in honour of Emerson Kemsies, head of the ornithology section of the University of Cincinnati, this southern red-tail subspecies is found in islands off the coast of Chiapas (in Mexico), Belize, and Nicaragua.

It is similar in appearance to the previously discussed hadropus, but displays rust-coloured barring only on the pendant feathers.

Buteo jamaicensis costaricensis (Ridgway)

As its name indicates, this subspecies is found mainly in Costa Rica (where it is known as the "Red-tailed Sparrow hawk" and is very common) although its zone of distribution also includes Nicaragua and Panama. This subspecies displays a wide variation in colouration.

The Caribbean

(Jamaica, Bahamas, Cuba, Dominican Republic)

Buteo jamaicensis jamaicensis (Gmelin)

Also known by the name of "Hispaniolan Red-tailed Hawk", specimens of this subspecies in Jamaica were first collected and classified by the German naturalist Gmelin, who gave their name to the species as a whole in 1788.

This subspecies is native to Jamaica, and is also found in the Dominican Republic, Puerto Rico, and on the Caribbean islands east of St. Kitts (with the exception of Bahamas and Cuba where they are replaced by the Buteo jamaicensis solitudinus (Johnsgard 1990).

They are small in size compared to other red-tails found in the Americas, with the males weighing an average of 700–800g (24.6–28.2 oz) and the females an average of 1.3 kg (45.8 oz).

They resemble the umbrinus of Florida, but have a rust coloured belly, darker colouration on the back, and less speckling. In addition, its thighs display rust-coloured barring. They are smaller than the red-tail subspecies found in Cuba.

Dennis Lorenz, a Swiss falconer, photographer and ornithologist who holds a Bachelors Degree in Zoology/Ornithology at Michigan State University, has worked with these Jamaican red-tails in the Dominican Republic, training them and hunting with them over a number of years. He told me that they are similar in colour to specimens of the calurus subspecies. He had a male that weighed 785g (27.7 oz), with a flying weight of 700g (24.7 oz). He also had a female that reached a maximum weight of 1,297g (45.7 oz) and that had a flying weight of 950g (33.5 oz). Both of these specimens were of average weight for their subspecies and sex.

Dennis has told me that the males have feet that are very powerful, considering their small size. A male that he trained was very fast and agile, which would be expected for its size. Its behaviour was completely identical to that of an accipiter or goshawk, and not at all like that of a red-tailed hawk. It could capture practically any bird in mid-flight without a problem. While not as agile as the male, the female was also very capable, and showed a high degree of aggression and skill in hunting mongoose, as well as chickens and wild game hens. On a number of occasions, it was able to capture cattle egrets after having been trained by Dennis specifically for this purpose.

According to Dennis, "**It always caught [the egret] on the ground but,**

The Red-tailed Hawk

after having suffered a series of pecking attacks, it no longer wanted to try to capture them". Once it went after a Yellow-crowned Night-Heron outside a palm grove, pursuing it for about 100 metres (109yrd) before giving up. Dennis says of the red-tail that it has a very calm character. It was very easy to work with it and to train it. "**It was incredibly gentle**". This is typically the case with red-tails when they have been correctly trained: they are very aggressive when they hunt, but never with the falconer.

Buteo jamaicensis solitudinus (Barbour)

Known as the "mountain sparrow hawk", this subspecies of Buteo jamaicensis is native to Cuba and the Bahamas. Those that live in Cuba are permanent residents and are distributed throughout the island of Cuba, in addition to Cayo Coco (which is on the archipelago of Sabana-Camagüey). In addition, these red-tails may be found on the *Isla de la Juventud*. It is somewhat similar to the umbrinus of Florida but smaller in size.

The solitudinus is found in abundance, residing mainly in forested areas, including wetland forests at sea level, at mid-range elevations, and at the highest elevations of Cuba, on the peaks of Cuba, Turquino and Suecia, which are part of the Sierra Maestra in the south-eastern region of the country.

This subspecies does not show any sexual variation in colouration, although it appears that, following the general pattern of red-tails, the females are larger than the males. The solitudinus adults have red tails with a subterminal brown-black band, a throat and upper neck that is reddish brown, a brown chest, and a whitish belly. The juveniles of the subspecies have a belly and chest that are whiter than those of the adults, with brown vertical streaks, and a tail that is brown-grey with dark brown barring.

Measurements taken using the Pie de Rey method, with an error probability of .05, of two specimens whose sex had not been determined, and that are part of the bird collection at the Institute of Ecology and Ecosystems, were as follows:

	Adult	**Juvenile**
Wing, folded	360 mm	362 mm
Tail	189 mm	192 mm
Beak, measured from the nare (culmen)	25.0 mm	23.3 mm
Tarsus	82.7 mm	83.9 mm

The solitudinus nests from January through July, constructing its nests in high trees, including the Royal Palm (Roystonea regia). It usually lays three cream-coloured eggs that have spots that are shaped like paintbrush strokes, and that are light brown or dark brown in colour. Measurements taken of a nest of three eggs, and which were deposited in the collection of birds of the Institute of Ecology and Ecosystems were as follows:

	Egg 1	Egg 2	Egg 3
Length	58.3	59.6	58.2
Width	43.4	43.5	42.4

The solitudinus has the habit of flying in circles at high altitudes close to the turkey vultures, using rising wind currents of hot air. When it does this, it looks as though it were similar in size to that bird, but the two species can be differentiated by the lack of a V shape of the wings of the solitudinus when it is gliding, and also because the solitudinus as a shorter and more open tail. When it is searching for its prey in the mountains, it glides slowly above the escarpments and, when it catches site of its quarry, it partially draws in its wings, and thrusts itself downward to capture them.

The solitudinus feeds primarily on reptiles and rats, although it can also eat birds and even Hutias (capromys pilorides).

Falconry is not practiced in Cuba, and birds of prey are generally not kept in captivity, although there are individual specimens of the mountain sparrow hawk in the collections Cuban zoos: for example, in the Havana zoo and at the National Zoo.

Buteo jamaicensis portoricensis (J.M. Iglesias)

This is the last of the subspecies of Buteo jamaicensis and is native to Puerto Rico. It is often confused with the variant of the subspecies borealis, which can also be found on the island. Portoricensis was discovered by the master falconer Juan Manuel Iglesias during his 15 years of research in Puerto Rico at a conservation centre for birds of prey.

Portoricensis is the smallest of all the red-tail subspecies, with the males weighing an average of 550g (19 oz) and the females and average of 750g (26 oz). This subspecies is easily differentiated from other red-tailed hawks, not only because of its smaller size but also because of its head, body and feet. It also displays behaviour that is different from that of other red-tails. Specifically, it tends to be calmer than other subspecies. It does not require a great deal of food in its diet: no more than 60–80g (2.1–2.8 oz) daily—equivalent to a chicken thigh. It should be borne in mind that all of the other subspecies of Buteo jamaicensis usually ingest large quantities of food, especially prior to reaching maturity. Some eat a quantity of food equivalent to an entire dove and even more, if they are hungry enough.

Juan Manuel Iglesias discovered this subspecies under very strange circumstances. A few years ago, another falconer on the island talked to him about a hawk that was just like the red-tail, but much smaller in size. Later, Juan Manuel came into possession of an adult male which weighed no more than 500g (18 oz). Later, he obtained another, similar male, but with juvenile plumage. He received the latter bird from La Entidad de la Red Caribeña de Varamientos. Iglesias was struck by the bird's small size—much smaller than usual for a Buteo jamaicensis. A few years later, in 1998, Iglesias received a chick that was only a week and a half old, and that was relatively small in comparison with a typical red-tail. At first, he thought that it might be a Buteo platypterus. However, it seemed a bit large to be a female of that species, and rather too small to be a male Buteo jamaicensis, given that it weighed less than 700g (24.6 oz). One day, at the age of 3 months and while it was moulting, it shed the central feather of its tail. Iglesias decided to save the feather in order to compare it with the feather of the adult. To his surprise, the new feather was a red-tail's feather, somewhat shorter than normal and with a single black stripe at the end of its tail that was finer than that of the typical red-tailed hawk. This subspecies generally prefers to feed on small reptiles, small birds, and rats.

Other American Hawks (Buteos)

I have included a brief description of the following American hawks, as some are also used in falconry and because they are very similar to the red-tail and have frequently been confused with certain subspecies of the Buteo jamaicensis.

Swainson's hawk (Buteo swainsoni) – Bonaparte 1838

This hawk is also known by the following names:

- *Swainson's Hawk*
- *Black Hawk*
- *Brown Hawk*
- *Grasshopper Hawk*

The Swainson's hawk, named for the English naturalist William Swainson (1789–1855), has also frequently been confused with specimens of the Buteo jamaicensis and, although these two hawks do bear a certain resemblance, they are also significantly different from one another in a number of ways.

The Swainson's hawk bears a strong resemblance to certain species of red-tails, in that it is large and yet slender, with very long and pointed wings. The Swainson's is the only hawk that, in its adult form, has dark-coloured feathers on the interior of its wings, but with lighter-coloured feathers on the exterior wings. Thus, the Swainson's exhibits wings of different forms. The average weight of these hawks is, for females, about 1,100 grams (38.8 oz) and for males, about 900 grams (31.7 oz). The wing span of the Swainson's is approximately 120–140 cm (47.2–55.1 in). Seem from the ground, it is possible to discern the dual tonality of the inner part of the wings. The Swainson's typically has a dark brown back, a dark bib, a white throat, and an abdominal area that is also white. Its plumage can be classified in three different phases, according to the age of the individual specimen:

- Adult (typically at two years of age, although some specimens do not display adult plumage until they are 3 or 4 years old),
- Sub-adult (one year of age)
- Juvenile (from hatching to one year)

At present, no subspecies of the Swainson's have been defined or officially recognised, although it is possible

2. Subspecies of the Buteo jamaicensis

to distinguish three distinct shades of colour among specimens of this species: a light colouration, a dark colouration, and a red or "rufous" coloration." There is little difference in appearance between the sexes in any of these variations. Consequently, it is not possible to determine sex based solely on plumage.

Individual specimens that exhibit dark colouration are relatively rare, although in some areas, such as California, they may constitute as much as 35% of the total population of Swainsons. Once again, the sexes display very similar plumage, although females are generally larger than males. There is, again variation in the plumage of adults, ranging from light shading to intermediate or rufous shading, to dark shading. The base colour of the interior parts of the adults, the plumage of the thighs, and of the coverts underneath the wings, may vary considerably, ranging from white, to reddish-brown, to dark brown, to black. The chest is usually darker than the abdomen (see photos at end of chapter). The tail is usually greyish-brown in colour, with streaks that are darker in colour, and with a sub-terminal band or line at the end of the tail that is wider, that has a white tip.

Individual specimens with white feathers and faded plumage have been observed, although these appear to be rare. There is no official record of albino specimens having been observed. They have often been confused with juvenile specimens of the red-tail, especially when they are perched since, like the latter, they display a faded but discernible "V" on their backs, their wings do not extend all the way back to their tails, they have white chests and dark abdominal bands, and they do not display a dark line above their eyes. In flight, the two species also closely resemble one another, except for the fact that red-tail juveniles have rounder wings, flight feathers that are lighter in colour, and display dark markings underneath the shoulders (these are referred to as "patagial marks" in field guides). Juvenile specimens of the Swainson, lack the abdominal band that is characteristic of the red-tail. The call of the Swainson's hawk is very similar to that of the red-tail, but is rather higher-pitched and somewhat weaker.

Habitat:

The Swainson's Hawk is frequently observed during the summer months in pasture and farming areas across a broad swathe of territory of western North America, from the Great Plains westward, and as far north as the Yukon Territory and Alaska. It is a hawk that thrives in arid and semi-arid zones, and that displays a clear preference for the wide open spaces of pasture areas and desert environments.

Hawks may often be seen sitting in meadows, or in open spaces. In this way, their preferred environments resemble those of certain red-tailed hawks. Swainsons may also be found in farming areas, as well as in places where there are an abundance of oak trees. In many instances, they follow the path of tractors in search of insects and rodents.

Distribution and Migration:

The Swainson's hawk is among the species of hawk that displays the highest levels of migratory behaviour, possibly because of its tendency to consume large quantities of insects. It is known for its long migrations, sometimes covering distances of nearly 20,000 km annually (about 12.427 miles). It is noteworthy that Swainsons are without exception a migratory species, and in this way are different to red-tailed hawks. The majority of Swainsons migrate to South American destinations—especially the Argentine pampas—where they spend the winter months. Like many northern red-tails, a minority of Swainsons migrate in groups to the south of Florida. There have also been reports of Swainsons migrating to Texas and California. They are rarely observed in the eastern United States during the autumn and spring months. The Swainsons generally migrate in medium-sized to very large groups. These groups are known as "kettles" in the United States. They frequently travel with turkey vultures and broad-winged hawks. During the migratory season, they can be seen in very large groups feeding on grasshoppers, their principle source of food, as well as certain species of rodent. When these species arrive at their winter destinations, they often can be observed feeding together on various types of insects. In Argentina, Swainsons have been observed feeding on locusts. This has led to them being known in parts of South America as "aguiluchos langosteros" (i.e., a Spanish term that literally means "eagle-like locust-feeding birds). The winter migration usually begins in August and ends in October. Due to their feeding habits, Swainson's hawks are very much appreciated by peasants, since they can be used to prevent and control plagues of insects.

Like other hawks, they tend to make use of thermal winds, especially during their migratory flights, in order to conserve energy, and when they utilize thermal wind currents in order to ascend, they are able to maintain their wings in the "V" position.

Diet:

On the basis of many studies, it has been determined that the Swainson's diet consists primarily of a combination of small mammals (e.g., moles, mice, ground squirrels, rats, and the occasional rabbit), and insects. It is also known that they consume a variety of birds, from the sparrow to the grouse, as well as small reptiles and snakes. At times other than breeding season, they usually prefer to eat insects rather than mammals, possibly because breading requires a diet that is richer in proteins. They may be observed in fields that are being ploughed, flying in the wake of tractors and hunting insects. In addition, Swainsons can observed feeding on insects—and on bats as well—when they are in flight during migratory periods. In this way, their feeding behaviour is similar to that of the Buteo jamaicensis. When there are fires, it is common to see Swainsons—as well as red-tails and other species of hawk, on the peripheries of the affected areas, hunting prey that are fleeing from the flames.

Reproduction:

Swainsons appear to be monogamous in their mating patterns, although it is hard to know just how long a period of time a couple spends together given that the species spends so much time migrating. Still, the belief persists among many ornithologists and falconry enthusiasts that Swainsons are lifetime monogamists. As is the case with most buteos, Swainsons reach sexual maturity at two years of age.

It appears to be the case that Swainsons often utilize the abandoned nests of crows, rooks, and other birds, and that they make certain modifications to these nests to suit their own purposes. In some instances, Swainsons construct new nests in trees, bushes, near pasture or farming areas, on the ground, or even on top of telephone poles (at heights ranging from one and a half metres to 30 metres (5ft to 98ft) above ground. The majorities of nests that are constructed in trees are less than five feet (1.5m) off the ground, and are generally located near pasture or farming areas. New nests are usually constructed in March or April. The process of constructing new nests or modifying old ones generally takes one to two weeks after the birds' arrive at their breeding grounds. Apparently it is the males that contribute the majority of the materials needed for construction of the nest, while the females decorate it. Both the male and the female of the couple line the nest with small branches of green leaves—something they continue to do after the eggs are hatched, and even after the chicks are born.

Female Swainsons usually lay two eggs. On occasion, they will lay as many as three. There is typically a two day interval between the laying of each egg. The female then usually devotes all of her time to sitting on the eggs until they are hatched, an incubation process that usually lasts between 33 and 36 days. In the meantime, the male hunts prey and provides food for both himself and the female. Sometimes, both the male and the female Swainson's share both hunting and incubating duties. The chicks usually hatch between March and July, and remain in the nest for about 30 days before permanently abandoning it at around 38–46 days of age.

Swainsons also share with red tails certain aggressiveness toward intruders who approach their nests, especially when they are incubating eggs or raising their newly hatched chicks. And, as is also the case with red tails, attacks against human beings have been known to occur for this very reason.

Swainsons will sometimes leave their nests permanently if they have been intruded upon. In fact in a magnificent video taken by George Robertson, and which he recorded in Arizona over the course of five months during which he conducted daily observations of a family of Swainsons, an unfortunate tragedy occurred when a nest that was located in a poplar, right in the middle of a cultivated field, was invaded one day by earth-moving equipment and trucks at a time when a recently hatched chick that was not yet ready to abandon the nest was forced to do so—just after his mother and father had done so—because of all of the surrounding commotion. The video was very instructive, and I would recommend it to anyone (a website address where you can order it may be found in the directory at the end of this book) even though it ends with the tragic death of the mother who had frantically but vainly searched for her chick. This event was in fact doubly tragic, since the Swainson's Hawk is currently classified as an endangered species.

In contrast to young red-tails, who are curious and who love to explore and investigate, there are some juvenile Swainsons that do not migrate with their parents during their first year of life—although the majority does migrate during their first winter (Brown, 1996, TPWD, 1997).

The annual moult of plumage of adults begins after the rearing of young or following reproduction, is suspended during the migrations, and begins anew once they arrive in territories where they will spend the winter. Juveniles usually undergo their first shedding prior to obtaining their sub-adult plumage. The subsequent sheddings take place in between March and the beginning of May, until they arrive at their summer destinations. This is the exact same pattern that is observed in the case of adults.

Swainson's hawks are often confused with specimens of the rough-legged hawk (Buteo lagopus) because these latter also have a white chest. However, in contrast to Swainsons, the legs of the rough-legged hawk are completely covered with feathers. On occasion, Swainsons have also been confused with light-coloured specimens of the meadow falcon (Falco mexicanus), when they are in repose. However these latter birds have darker eyes, the ends of their wings do not reach back all the way to their tails, and juveniles of the species tend to be reminiscent of saker falcons, since they do closely resemble them.

The Swainson's hawk is a species of hawk that is an object of considerable concern, since its population has steadily declined since the 1940s in certain areas, such as California. In some places, the population of Swainsons has dropped by more than 90% (Bloom et. Al., 1985; Schlorff, 1985) and it continues to be listed as an endangered species by The United States Fish and Wildlife Service.

Between the years 1995–1996, almost 6,000 Swainson's hawks died in the Argentine pampas, mainly as a result of a pesticide used to control a plague of crickets.

The web page of the organization Friends of the Swainson's hawk (based in Sacramento, California) contains more detailed information regarding this bird, and also contains helpful suggestions aimed at avoiding the kinds of unfortunate events described above. The directory at the end of this book contains contact information of this organisation. Many wonderful photographs taken by George Robertson may also be found there.

Broad-winged hawk (Buteo platypterus) – Vieillot 1812

This hawk is also known by the following names:

- *Broad-winged hawk*
- *Broad-winged buzzard*

2. Subspecies of the Buteo jamaicensis

This hawk has sometimes been confused with specimens of Buteo jamaicensis, despite its small size, which is approximately identical to that of a crow. The broad-winged hawk is the smallest of the North American buteos, and is similar in appearance to juvenile specimens of both the red-shouldered hawk (Buteo lineatus) and Cooper's hawks (Accipiter cooperi). It can be found in the summer, mainly in northern and southern forests throughout the continent. Broad-winged hawks also vary in colour, although darker coloured specimens are rare, and have only been observed in Canada. In comparison to other hawks, the tips of their wings are very pointed in appearance. The males and females of this species are very similar in appearance, with females generally being somewhat larger in size. The adult plumage of broad-winged hawks is different to their juvenile plumage.

This species can easily be identified by its small size and by its distinctive tail, which in adults as two or three alternating broad black and white stripes while in juveniles there are 5 alternating black and brown stripes. Another characteristic of this species its beak, which is entirely black except for a tiny blue point on its lower jaw. It has short tarsi, and wings that are rounded, sharp, and wide. Its plumage varies with the age of the individual specimen. Like other hawks, it can be divided into the following: adults, sub-adults, and juveniles. Like other birds of prey, the juveniles of the species have longer feathers and tails although the width of the wings tends to be constant throughout the life span of the species.

Lighter-coloured and darker coloured variants are only found in the subspecies (Buteo platypterus platypterus) that resides in North America. Its wing span is 81–91 cm (32–36 in).

A total of six subspecies of the broad-winged hawk have been identified, although B.p. platypterus is the only one that has been observed in North America. The other five subspecies, found in the Caribbean, are: B.p. cubanensis (Cuba); B.p. brunnescens (Puerto Rico); B.p. rivieri (Dominica, St. Lucia, and Martinique) and, finally, B.p. antillarum (Granada, St. Vicente, and Trinidad and Tobago).

Habitat:

This bird of prey resides in mixed forests in an area extending from central and south Canada to the Eastern US. This species tends to make use of very dense areas of forests that are not even visited by red-tailed or red-shouldered hawks. Broad-winged hawks can frequently be seen in clearings that are created by the construction of highways, or in swamplands.

Distribution and Migration:

Despite the fact that this is a migratory species, its geographical distribution of this species is very similar to that of the red-tailed hawk. As mentioned above, the North American subspecies is known as the Buteo platypterus platypterus. There have been an additional two subspecies who have been officially recognized: B.p. brunnescens (Danforth and Smith), which is native to Puerto Rico, and which is in danger of extinction, and B.p. Cubanensis (Burns), which is found only in Cuba, and which flourishes in that island nation. This is one of the most common birds of prey in North America.

This species spends its winters in parts of Central and South America, including Peru and Brazil. Occasionally specimens are observed in Florida, along the coastal area of southeast Texas, and in the northern Mississippi delta. Specimens observed in these areas of the United States have been, for the most part, juveniles.

As is the case with the Swainson's hawk, this species is completely migratory, with the majority migrating to Central and South America. Some specimens have been observed as far south as Chile and Argentina. It is only during migrations that these birds may be said to be gregarious, since they are generally rather solitary, although they are also rather gentle (tame). Other specimens prefer to migrate to the Gulf Coast of Florida, along with several subspecies of red-tailed hawk, such as the northern borealis or kriderii and other hawks, such as the red-shouldered hawk.

Every once in a while, an individual specimen is observed in the western US. Melanistic forms of the species are also quite rare, but have been observed in both the western Canadian province of Alberta as well as in the Great Plains. Broad-winged hawks generally construct their nests either within the forests where they preserve to reside or near bodies of water.

Diet:
The diet of the broad-winged hawk consists mainly of insects, but also includes locusts and rodents, as well as a number of amphibians and reptiles. This hawk captures its prey both on the ground and in trees. It does not feed on carrion and sometimes will hunt from atop poles. At times, it will utilise the technique of "still-hunting", although from much lower heights than the red-tailed hawk. At times, the broad-winged hawk will fast during its migrations.

Its size is rather small in comparison with some subspecies of red-tailed hawk, with males and females weighing an average of 350 g and 450 g respectively. The male of the species are sometimes confused with the Puerto Rican male red-tailed hawk (Buteo jamaicensis portoricensis) because they are similar in size.

Reproduction:
The courtship behaviour of the broad-winged hawk is very similar to that of other buteos, and it is believed that this species is monogamous. However, as is the case with the Swainson's hawk, it is difficult to ascertain whether this is true due to the migratory nature of the species. Broad-winged hawks will often make their homes in nests abandoned by other birds of prey, or by crows or squirrels, renovating the nests with leaves and grasses and tidying them up. At times, they also will construct their own nests from scratch, although they are not generally known as good nest-builders. Broad-winged hawks will re-use their nests for a year or two at most: they never use the same nest for three or four years in a row. And these nests are usually located near the tree trunk, near the fork of the tree. Less

commonly, they will make their nests higher up in a tree. Broad-wings use many different kinds of trees for nesting, although those most commonly utilised are birch, poplar, and maple trees. The females generally lay two or three eggs, although they may lay as many as four. It is the female alone who incubates these eggs during a period lasting between 25 and 31 days, with the male doing the hunting and providing food for both himself and his mate—except when the female herself decides to go hunting. When the chicks are hatched, they are fed by the female for the first 29 to 30 days of their lives, at which time they begin to leave the nest. As is the case with other hawks, the parents continue to help feed their chicks outside of the nest, until they reach the age of just over two months.

Once again, they also just as aggressive as red-tails during breeding season and have been known to attack humans who got to close to their nest or who made them feel threatened.

Ferruginous hawk (Buteo regalis) – Gray (1844)

This hawk is also known by the following names:

- *Ferruginous rough-legged*
- *Squirrel hawk*
- *Eagle hawk*
- *Gopher hawk*

The popular name of this hawk comes from the Latin word *ferrugo*, which refers to the "rust colour" of this species. Its scientific name in Latin means "royal hawk", and refers to the large size of this bird, which ranges from 900 g to 2,030 g (32–72 oz). Its wing span is generally between 135 and 168 cm (53–66 in). There are specimens that are even exceed the outer limits of the ranges stated here, but there are also males that are even smaller in size than 900 g (32 oz).

Ferruginous hawks display distinct colours of plumage, with lighter, reddish (or "rufous" tones being more prevalent than darker tones. Males and females are very similar in their plumage, but females are generally larger in size, and their feet are a great deal larger than those of males. A distinctive characteristic of this species is their large black beak with a blue tip on the lower jaw, and their elongated cere (i.e., the soft part of the upper jaw). Also distinguishing them from other species are their feathered tarsi and, as mentioned previously, their large size. The juvenile plumage of this bird differs from that of adults of the species, lacking the reddish or rust colour on the upper parts of the body, on the feet, and on the tail. Among the species as a whole, it is possible to distinguish among four distinct types, according to the colour of their plumage: light, intermediate (or reddish), dark-intermediate, and dark. Among light-coloured specimens, there may be slight variations in plumage. Plumage also varies with age, with juveniles passing through three phases (i.e., light, intermediate, and dark) during which their plumage grows relatively darker in colour. In contrast to other large-sized hawks, the ferruginous hawk does not have a distinct sub-adult plumage. No subspecies of this hawk have been officially recognized.

This species is very robust, with long and broad wings with rounded tips, large heads, and large beaks (with their mouths opening up to a diameter of 42 mm or 1.7 in). These large beaks pose a problem when these birds are used in falconry, because they enable ferruginous hawks to devour large quantities of foods in just a few seconds (and therefore a large part of the prey may be devoured by the ferruginous prior before being delivered to the hunter).

Habitat:

The ferruginous hawk lives in arid, open spaces, dry prairie lands, and intermountain regions of the western US and Canada. It seems to prefer uncultivated prairie lands, and it is able to survive in such environments even when there are no trees.

Distribution and Migration:

The ferruginous hawk is the largest and heaviest species of buteo, and makes its home in the states of the Great Plains, and the West and Southwest, of the United States and Canada. It is sometimes confused with certain subspecies of the red-tailed hawk, especially the Buteo jamaicensis kriderii. In winter, some ferruginous hawks migrate to the South, East, and West. There are some members of the species that reside in southern zones, although some of these will also migrate in search of new hunting grounds.

The ferruginous hawk displays a high degree of adaptability to open and uncultivated spaces, and to the semi-arid zones of the Great Plains. It will hunt from and nest in tall trees near rivers or lakes, but can also thrive in open spaces that have few or no trees.

Diet:

The feeding habits of the ferruginous hawk are very similar to those of red-tails, although the two species generally do not compete with one another over the same prey, since they hunt at different times of day: hunting very early in the early morning hours or shortly before sunset. It primarily feeds on mammals, especially cottontail rabbits, hares (Lepus californicus or Lepus townsendii), squirrels (including the Sperophilus richardsonii), wild prairie dogs, and rodents that make their homes in pasture areas. Ferruginous hawks pursue hares, grouse, partridges and pheasants at low altitudes.

The hunting strategy of ferruginous hawks appears to be similar to that of red-tails. They start from an observation point atop a pole, circling their prey from above, rapidly covering long distances and flying near ground level in pursuit. The ferruginous are also known for hunting cooperatively in pairs. Their prey consists primarily of ground squirrels and hares, although they occasionally feed on mammals and poultry. They employ four different types of hunting strategies:

1. Attacking prey at ground level.

2. Pursuing prey from the air, at a height no greater than 30 metres (98ft).

3. Pursuing prey from the air at heights greater than 100 metres (328ft).

4. Hunting from posts or self-created "guard posts"—still-hunting, just like red-tails, followed by a maximum pursuit of 100 metres (109 yrd).

The first and second of the strategies listed above appear to enjoy the greatest success.

Reproduction:

As is the case with red-tails and many other species of hawk, ferruginous hawks appear to have monogamous mating patterns. Like nearly all buteos, they reach sexual maturity at the age of about two years. The courtship patterns of the ferruginous appear to diverge from that of most other hawks, in that the male does not attempt to attract a mate by performing particular "flying stunts". Males obtain their mates after a prolonged period of time, and have rarely been observed feeding the females beforehand. It appears that the bonds of the couple are forged through the creation of the nest. The behaviour of the male during the breeding period is in effect that of simulating the construction of a nest. After thus attracting the attention of a female, the two copulate (without any associated flying). After copulation, the male hunts and obtains food for the female.

As for the nests themselves, these tend to be built in the same areas as those of red-tails, which indicates that there is possible competition between the two species as regards both nest location and prey. The ferruginous have a curious habit—one they share with some red-tails—of constructing more than one nest at a time. It is thought that they may, after doing so, switch nests from one year to the next. As is the case with many red-tails also, when they lose eggs prior to hatching at the beginning of breeding season, they will often immediately move to one of the nests that they have previously constructed to begin the process anew. The vast majority of these nests are constructed in trees, although such nests usually produce the worst results. The remainder of nests are constructed either on the ground or on rocks in hilly areas. The common element in each of these nest locations is an open view of the surroundings and a location that appears to be far away from the possibility of any human intrusion. Both members of the couple construct and arrange the nest, although it is the male who collects most of the material from the ground or from bushes, and the female who spends most of her time arranging the nest and shaping its foundation. If a formerly used nest has been chosen by the couple, it may be one that has been a rather large one that has served as the home of either red-tails or Swainsons. Sometimes, ferruginous hawks will build their nests on top of nests that have been previously used by magpies. Whatever the case may be, ferruginous nests are generally quite large, with diameters ranging between 61 and 107 cm (24–42 in). These nests are usually made of twigs (like those of red-tails), but often they will add bones, leaves and grass, bark, and even the manure of cows, horses or buffaloes in the interior of the nest.

The number of eggs that are laid by the female varies greatly. As is the case with other birds of prey, this depends on the amount of food available as well as the type and location of the nest. It appears that eggs laid in nests in trees are smaller, and that they also carry a higher risk of mortality (possibly because of the eggs or the chicks falling out of the nest) than those laid in nests constructed on the ground. Generally speaking ferruginous females lay between two and four eggs during each breeding season, with an interval about two days between the laying of each egg. The incubation period between the laying and the hatching of the eggs is about 33 days following the laying of the first egg. Both the male and the female of the couple share the task of incubating the eggs, although the latter spends more time on this task. Chicks abandon the nest 38 to 50 days following hatching, and it is generally the males, that are smaller in size, which abandon the nest first—at times up to ten days before female chicks.

It is commonly believed that the rough-legged buzzard (Buteo lagopus) and the ferruginous hawk are closely related, since they both have not only feathered tarsi—as do eagles—but also broad beaks (and a consequent ability to swallow large quantities of food). However, the ferruginous hawk hunts much larger prey than the rough-legged hawk.

The ferruginous hawk is not a particularly prolific species and, during the years 1971–1981, it was on the Audubon Society's list as a species in decline. It was also listed as a species about which there was "great concern" by the US Fish and Wildlife Service from 1982–1986. In some American states, it remains either a threatened or an endangered species. It is currently considered a threatened species in Canada and is now once again being considered for inclusion on the list of threatened species by the US Department of the Interior. Because of this precarious situation, platforms have been constructed in American states and Canadian provinces in order to enable ferruginous hawks to construct nests and rear their young in the most propitious conditions possible. The government of the United States actually has provided subsidies and other forms of assistance to farmers who allow their lands to revert to their natural state, or who plant grass. Electricity poles have been modified in areas of high concentration of birds of prey in order to decrease incidents of electrocution. In addition to these efforts on the part of federal governments there have been initiatives on the part of both local governments and non-profit organizations to provide nesting and feeding areas for ferruginous hawks.

Rough-legged hawk (Buteo lagopus) – Pontoppidan 1763

This hawk is also known by the following names:

- *American rough-legged*
- *Rough-legged buzzard*

2. Subspecies of the Buteo jamaicensis

In comparison to other buteos, the rough-legged buzzard has a small beak and hands. It also has feet that are completely covered with feathers up to the tarsi. Females are larger than males. As for its size, it is larger than the common buzzard (Buteo buteo). The male and female of the rough-legged buzzard are very similar in appearance.

The species displays two different types of plumage: light-coloured and dark coloured. The latter are more prevalent than the former—and to a far greater extent than in any other species of hawk. This is especially the case in eastern zones, and thus the relative pattern of distribution of the two types is exactly the opposite of that of red-tails. The sexes present only a slight difference in plumage. Females are larger in size than males. It should be noted that females also present an intermediate phase of coloration.

Three different subspecies of the rough-legged buzzard are known. The most common of these is the Buteo lagopus sancti-johannis, is the one native to North America. The B.l. lagopus breeds in tundra environments, as well as in European forests, in an area extending eastward from Norway to Russia. The European variants of the bird found in this zone migrate to coastal areas of the United Kingdom for the winter, while those native to France winter either in northern Italy or Turkey. Another of the subspecies, the B.l. kamtschatkensis, breeds in the tundra of the Siberian Urals in a zone extending eastward and southward to the Bering Straits and the Kuril Islands respectively. This subspecies winters in areas along the Caspian Sea, northern China, Korea and Japan. The kamtchatkensis is the largest of the rough-legged subspecies, and is very similar in appearance to its North American cousin. This subspecies does not have a dark-coloured variant.

The rough-legged buzzard has often been confused with several subspecies of red-tail hawk, such as the Buteo jamaicensis harlani. These two types of hawk in fact often do interbreed, producing hybrids that are even darker than dark coloured ferruginous hawks.

This species is average in size in comparison to other hawks, ranging from 745 g to 1,380 g (26–49 oz). The average weight of males is nearly one kilogram (35.2 oz), and the average weight of females is nearly 1.3 Kg (45.8 oz). The wing-span of the species ranges from 122–143 cm (48–56 in). The main differences between this hawk and the red-tail are that the latter do not have feathers on their legs. But they are similar to red-tails in that they present a variation in plumage and bear a certain physical resemblance to one another. Yet rough-legged buzzards seem to more closely resemble ferruginous hawks (Buteo regalis) of the eastern United States, and are also closely related to an Asian species of hawk, the Buteo hemilasius. The rough-legged hawk has long, broad, and rounded wings. Its tarsi and cere are of an intense yellow colour. On occasion, these birds can be observed hovering in flight.

Light-coloured variant

The sub-terminal band of the tail resembles that of the red-tailed hawk, in that it is wide and very dark, with the upper portion of the tail having a dark brown colour with speckling. Light-coloured females have an abdominal area that is darker than that of males, while males have a rather darker chest than females.

Light-coloured juveniles are similar in appearance to adult females, with an abdominal area that is black in colour. The upper areas of their primary coverts are white, while the subterminal bands of their tails are dark, over a lighter barred base.

Dark coloured variant

The dark coloured variant of rough-legged buzzards have very dark plumage, with the flight feathers that are streaked as well as lighter in colour than the interior wing feathers. The ends of the feathers have very dark edges. Similar to adult red-tails, they have a wide and dark sub-terminal band on their tails. One characteristic that distinguishes the female is a white base on the tail. The male has a dark tail with narrow, lighter coloured streaks. There is a high degree of similarity between dark coloured adults and juveniles, although the latter are often darker brown in colour and have a less pronounced sub-terminal band on their tails. The tails of juveniles is generally lighter in colour than that of adults, with sparse and somewhat less pronounced streaks of a darker colour. Juveniles also have a white patch on their primary wings (feathers).

Habitat:

The rough-legged buzzard tends to make its home in cleared areas, tundra zones, or trees that are on the border of the tundra or the forest. In winter, or during migrations, it also tends to prefer open spaces, including pasture areas, farm areas, or plateaus. In the Americas, they look for zones similar to arctic tundra (i.e., the Great Plains), prairies, airports, coastal swamplands and open areas that afford excellent visibility, and where they are able to hunt rodents. In the winter, they tend to form communal night-time roosting areas.

Distribution and Migration:

Geographically, rough-legged buzzards can be found in the northern portion of North America: i.e., in Alaska, throughout northern Canada, areas that feature the wide-open spaces that they tend to prefer. These buzzards are migratory, and often winter in southern Canada and the northern United States where they also make their homes in open areas, especially farms, swamplands and airports. Although this species is widely distributed across the arctic regions of North America, Europe, and Asia, there is relatively little information available about it. The subspecies of the rough-legged hawk that may be found in the United States is the Buteo lagopus sancti-johannis (Gmelin).

Diet:

Rough-legged buzzards feed almost exclusively on small- and medium-sized mammals, and frequently eat carrion, especially animals that succumb to the harsh winters of the lands where they make their homes. According to a number of studies, it appears to be the case that they prefer small voles and lemmings to other mammals. Birds, on the other hand, are a rather small part of their regular diet. Among the birds they do feed on, they seem to show a marked preference for pheasants and grey partridges, and for an occasional domesticated chicken or hen.

They also hunt in a similar way to red-tails, but usually from shorter posts or closer to the ground. They have also been seen to appear to be hovering in light winds.

Reproduction:
Although it is known that the rough-legged buzzard begins to reproduce at the age of two years, there is little additional information about its reproductive patterns. It appears that these birds have monogamous mating patterns, since they have often been observed arriving in couples at their breeding grounds. Their courtship habits of the North American subspecies have not been studied, but it is widely assumed that they are similar to those of other American hawks. It is known that they tend to use the same nests year after year, gradually enlarging them with each passing year. To a large extent, they tend to construct their nests on cliffs or in similar places. In this way, they are similar to falcons, although rough-legged buzzards prefer to build their nests in areas that afford at least some cover. These hawks use the same nesting areas as other birds of prey that make their homes in Arctic clearings, such as gyrfalcons, peregrine falcons, and other birds, although they are more plentiful in number than these other species.

Where there is a shortage of trees, they will use platforms, clearings, or cliffs, and may even build their nests on rocks that are located on hills, just as ferruginous hawks do. They have a preference for "covered" nests and when they do make use of old nests, these grow in size year after year.

Rough-legged hens lay a variable number of eggs, the number of which may range between two and seven, depending on diet. The average number of eggs laid during a single breeding season is four. In captivity, the interval between the laying of eggs is usually two days, and the total period of incubation generally lasts between 31 and 37 days. Chicks remain in the nest for approximately 40 days. As is the case with many other buteos, male chicks generally abandon the nest before female chicks.

Like with other buteos, rough-legged Hawks have a variety of plumage colours that range from light to dark. Also, the male, female, and juvenile birds have slightly different plumage patterns and like other hawks, females usually are larger.

Red-shouldered hawk (Buteo lineatus) - Gmelin 1758

Other names for this hawk:
- *Florida red-shouldered hawk (alleni)*
- *Northern red Shouldered Hawk (lineatus)*

- *Red-bellied hawk (elegans)*
- *Texas red-shouldered hawk (texanus)*

The red-shouldered hawk is very large in size, with a rather long tail and broad wings. The male and female of this species resemble one another in appearance. Their beaks are short, curved, and dark in colour. The base of their primary wing coverts is white and translucent. Their tails are brown, with white bands. There are differences in appearance between adults and juveniles.

With a weight of weight of about 700 g (25 oz) and of medium height (43–61 cm / 17–24 in) in comparison to other species of hawk, the colour of adult red-shouldered hawks ranges from light brown to dark-brown (depending on the subspecies). Some subspecies also feature white speckling on the chest, and both the chest and wings are streaked. Rufous highlights may be found on any area of the red-shoulder's body, but have been observed most frequently on the shoulders (this is what gives the hawk its name). Their tail is black, with four narrow white bands. The chest, abdomen, and interior wings have a rust colour. These hawks have longer tails than red-tailed hawks and, like red-tails, their colour tends to darken with age.

Juveniles of the species are brown in their upper bodies, with very little reddish colouration. The proximal areas of their bodies generally have a cream colour, with patches or specks of red. The colour of juveniles' tails is greyish brown with narrow bands of light brown.

They fly in a manner similar to goshawks, rapidly beating their wings three to five times before coasting. They do not hover over their prey.

Recognised subspecies (Johnsgard 1990):

- *B.l. lineatus (Gmelin) – Breeds in the Eastern US and Canada, in southern Texas, and in Nuevo León, Mexico.*
- *B.l. alleni (Ridgway) – Resides in a zone extending from Texas and Oklahoma eastward to South Carolina and Florida (lighter in colour).*
- *B.l. extimus (Bangs) – Resides in the extreme south of Florida and in the Florida Keys (lighter in colour).*
- *B.l. texanus (Bishop) – Resides in a zone extending from southern Texas southward to Veracruz and the Federal District of Mexico. (has brighter red colouration, like the calurus red-tailed hawk.*
- *B.l. elegans (Cassin) – Resides in California, western Oregon, and northern Baja California (also redder in colour).*

Distribution and Migration:

The red-shouldered hawk reproduces in the Great Plains, from southern Canada to the Gulf Coast, as well as in central Mexico. In California, an isolated population of these hawks has also been found. Red-shouldered hawks that reside most of the year in northern climates are migratory—as is the case with red-tailed hawks.

Habitat:

This bird of prey may be found in humid areas of the forest, in forested areas near rivers, alongside swamplands and sparse pine forests, and in similar environments. Red-shouldered hawks almost always construct their nests along a body of water, whether it is a swamp, puddle, or pond. They tend to reside in forests of denser growth than those preferred by red-tails. They have also recently been observed in residential areas, although it is difficult to observe them perched on the branches of trees.

Diet:

Red-shouldered hawks feed primarily on small- to medium-sized mammals. The largest mammals among their prey are rabbits and squirrels. They also eat small birds, snakes, and medium-sized amphibians and reptiles, locusts, and large insects. They usually hunt alone, from observation posts located near bodies of water.

Some of their hunting techniques resemble those of red-tails and other hawks, since they make use of observation posts or trees as launching points for diving attacks against their prey. Alternatively, they may begin diving toward their prey from mid-air.

Reproduction:

Red-shouldered hawks begin reproducing at two years of age, with females of the species generally laying two or three eggs during a single breeding cycle, either in April or May. Eggs are incubated for 28 to 32 days by the female, while the male provides food for himself and his partner. Nests are constructed in oak and pine trees, or in other large trees, generally at a height of 10 to 20 metres (33–66 ft) above ground and in an area protected by the tree's leaves. Nests are constructed with sticks, twigs, bark, and moss. Once construction of the nest is finished, the male places leafy branches around its exterior as a means of marking his territory. Once chicks have been hatched, both the male and the female share in hunting duties. The female usually captures slower and bigger prey, while the male hunts smaller and swifter animals. Like red-tails, juvenile specimens of this species usually abandon the nest at the age of about 45 days.

At times other than breeding seasons, the red-shouldered hawk is a solitary creature. At all times and, again, like the red-tailed hawk, it is a species that at all times exhibits a high degree of territorial behaviour. Males may exhibit threatening behaviour toward other males of the species—and in fact toward any intruders—that infringe upon their territory.

Because they red-shouldered hawks often bear a strong resemblance to juvenile specimens of red-tailed hawks, they have often been confused with this latter species, even though red-tails tend to have narrower wings and longer tails.

The Red-tailed Hawk

Common North American Buteos

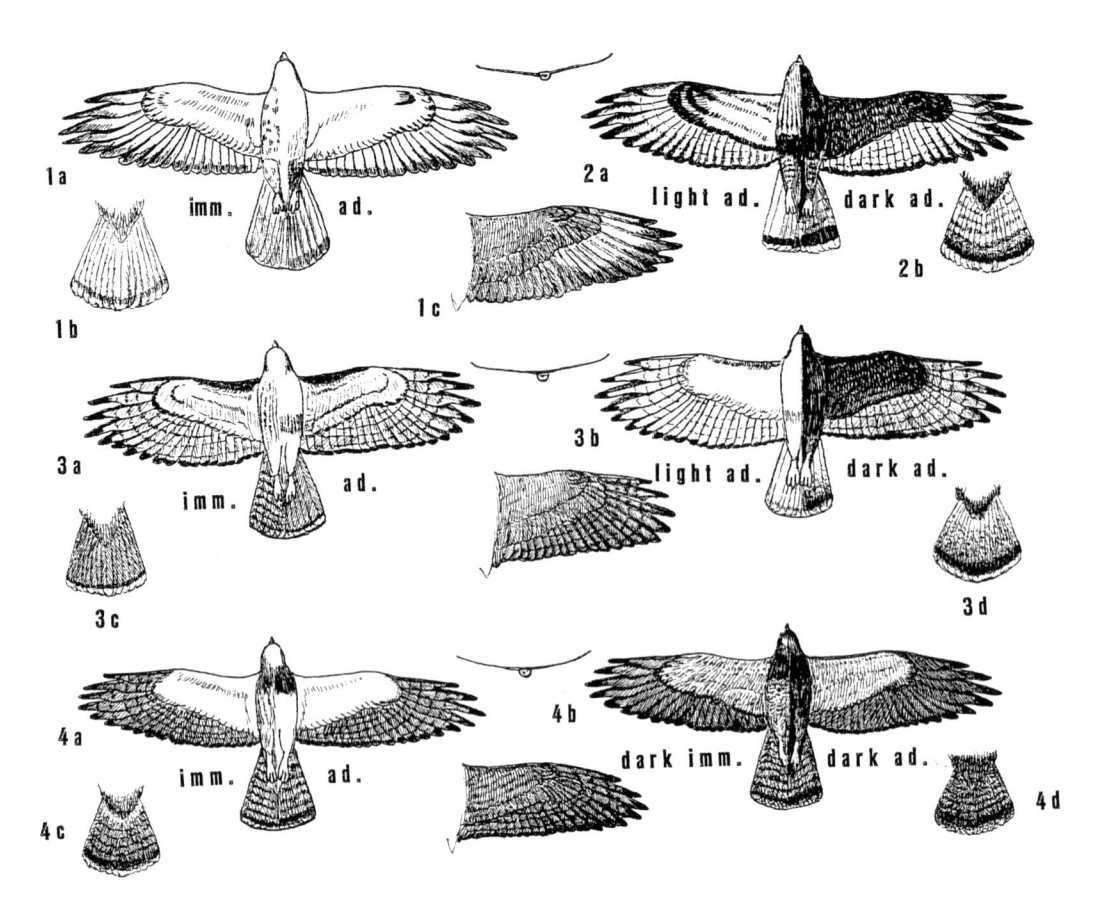

Field drawings of Dr. Paul A. Johnsgard for identifying the most common buteos that can be found in North America

Number 1 shows a ferruginous hawk (1ª; adult tail and wings from above, 1b and c).

Number 2 shows a rough-legged hawk (2ª: tail of adult male seen from above 2b).

Number 3 shows a red-tailed hawk (typical phase 3ª, kriderii and harlani 3b, adult tail seen from above, typical phase 3c, Harlani 3d, wing from above 3e).

Number 4 shows a Swainson's hawk (typical colouring 4ª, dark phase 4b; typical adult tails and tails of dark phases, 4c y 4d: wing seen from above 4e).

2. Subspecies of the Buteo jamaicensis

*Close-up of female Buteo jamaicensis, possibly an eastern subspecies.
Photo courtesy of Fernando Flores.*

The Red-tailed Hawk

Adult albino Buteo jamaicensis. Albino hawks with light coloured eyes usually have poorer eyesight than those individuals with normal coloured eyes. Albinos are also more susceptible to disease and seem to have a weaker immune system. These rare individuals can be found in the centre and east of the US. Albino hawks are practically unknown in the west.
Photo by Charlie Kaiser.

Juvenile ferruginous hawk, the biggest of all North American buteos.
Photo by Charlie Kaiser.

2. Subspecies of the Buteo jamaicensis

Light phase adult red-tail, calurus subspecies, Lamar, CO, USA. These red-tails are amongst the darkest of light phase individuals of this subspecies. The belly band and pendant feathers usually appear with barring, this being a typical feature of the "calurus" subspecies. Photo by Brian K. Wheeler.

Florida red-tail, also known as the Buteo jamaicensis umbrinus. Resident all year round in Florida, where they replace the borealis subspecies, though juveniles are very similar to these. Adults, however, are darker (especially the head and top parts) and look more like western red-tails. Photo by Brian K. Wheeler.

The Red-tailed Hawk

Buteo jamaicensis harlani, intermediate phase. These individuals have a dark head with a partially white frown and brow. Throat is usually white while the breast, belly and sides are predominantly speckled in white with black and dark brown barring on the feathers. Photo by Brian K. Wheeler.

Adult Buteo jamaicensis harlani, moderately dark phase. These individuals appear with a considerable amount of white speckling both on the head and chest which has brown and black barring and are very similar to the dark common buzzard individuals. This subspecies is the only one to have a greyish tail instead of the intense red hue that is typical of this species. Photo by Brian K. Wheeler.

2. Subspecies of the Buteo jamaicensis

Close-ups of "Beta", female Buteo jamaicensis calurus performing in a falconry display at Safari Madrid, Spain. This is the typical western red-tail, darker and more rufous than eastern subspecies, and of a medium size amongst red-tails. It is very similar to the common buzzard that we know in Europe (Buteo buteo), and has a shorter tail with broader and longer wings than eastern subspecies. In Spain, the UK and most parts of Europe, these are the red-tails that we are used to seeing, as most captive-bred individuals belong to this subspecies. Below, we can see a little closer the feet of these red-tails which although quite powerful are not amongst the biggest or most powerful amongst red-tails (this being important when hunting squirrel or hare), allow both males and females to be perfectly able to hunt hare. Due to their size, however, I would always recommend a female of this subspecies for such quarry. Photos by author.

The Red-tailed Hawk

Juvenile Buteo jamaicensis borealis; a typical eastern red-tail, in Lamar, CO, USA (November). Photo by Brian K Wheeler.

Juvenile borealis individual in flight, Duluth MN, USA (October). Juveniles of this subspecies appear as very white when seen from underneath, although they are heavily marked in their abdominal region and usually have a bib which is sometimes outlined by a collar or is open just like in the above photo. The pendant feathers can be barred in some individuals and the back is usually of a dark brown colour with abundant white speckling, just like in adults; juveniles are very similar to these with one big difference which are lighter markings in the abdominal region of adults. Photo by Brian K Wheeler.

2. Subspecies of the Buteo jamaicensis

Adult Buteo jamaicensis kriderii, Houston, TX; USA. This subspecies has the most white in its plumage while not presenting signs of albinism. The tail of these individuals is still red, but it is a much lighter shade than in other red-tails. Photo by Brian K. Wheeler.

Adult kriderii red-tail in flight, Tulsa, OK, USA, (November). Photo by Brian K. Wheeler.

The Red-tailed Hawk

Buteo jamaicensis fuertesi. This is the red-tail with the biggest wingspan of all and we can find it mainly in dry, desert zones especially in the areas of Texas, Baja California and Mexico. Photo by Brian K. Wheeler.

2. Subspecies of the Buteo jamaicensis

Kira, female Buteo jamaicensis jamaicensis by Yannick Lorenz. These are the individuals that we will see in Jamaica and Dominican republic. Below, to the left, (photo by Dennis Lorenz) we can see the typical feature in the tail of southern and western individuals; in addition to the dark subterminal band of the tail, we can observe other dark bands, much narrower and incomplete, throughout the tail. Bottom left photo by Eladio Fernández.

The Red-tailed Hawk

*Close-up of an adorable and curious Swainson's hawk chick (top). At times, these hawks have been confused with light-phased prairie falcons, also known as "American sakers" (Falco mexicanus) particularly when perched. Juveniles can also greatly resemble saker falcons (Falco cherrug), although obviously, there is a difference in size.
Photos by George Robertson.*

2. Subspecies of the Buteo jamaicensis

Swainson's hawk, adult, at sunset. Photo by George Robertson.

Adult broad-winged hawk (Buteo platypterus), Kit Carson, CO, USA (September). Photo by Brian K. Wheeler.

The Red-tailed Hawk

Adult female red-shouldered hawk, alleni subspecies, also known as "southern" red-shouldered hawk (February). Photo by Brian K. Wheeler.

Male rough-legged hawk (Buteo lagopus), light-phase individual, Amarillo, TX; USA (November). Photo by Brian K Wheeler.

2. Subspecies of the Buteo jamaicensis

The Red-tailed Hawk

Pale Male flying over the moon. Photo by Lincoln Karim.

3. The World of the Red-tailed Hawk

Although in Europe, the red-tailed hawk continues to be a largely unknown entity, in other countries, like the United States for example, the species has been used in falconry for at least the past forty years.

In this chapter, I will describe the situation of the red-tailed hawk in some of the countries where it is found, so that you can see, as I have, that red-tailed hawk has been known and appreciated for a very long time in countries that rank among the world leaders in the practice of falconry, and that it is indeed a bird that is capable of excelling in this sport. It just needs to be given a chance.

United States

In the United States, falconry is a recognized sport that is regulated by federal legislation, although there continues to be a number of states where it is outlawed (e.g., Hawaii). The sport is also currently illegal in the US territory of Puerto Rico. The practice of falconry varies from one US state to another since, in addition to the federal regulations, each state also imposes its own particular state regulations. This is similar to the situation that prevails in Spain with respect to the Autonomous Communities of that nation, where some states are more restrictive than others and where, in some places, the practice of the sport is forbidden altogether.

It is generally true that, in the states where it can be legally practiced, there are certain common requirements, such as a minimum age of 14 for falconers' apprentices. There is a written exam on the biology of birds of prey, natural history, falconry, and legislation and regulations related to the sport that must be passed before one can embark upon an apprenticeship. This exam is administered by each individual state, and it is also required that each apprentice have a designated guide, an experienced falconer, that will serve as his or her tutor and mentor during a period of two years, at the end of which time the apprentice will be eligible to obtain a license to practice falconry independently. A government official[1] is also authorized to carry out an inspection of all falconry equipment that is to be used, as well as the installations where birds are kept.

The falconer must comply with each and every one of these requirements before being able to begin to work with the bird of his or her choice. In the US, the only two birds that can legally be used in falconry by beginners are the red-tailed hawk and the passage kestrel (except for the state of Alaska, where the goshawk may also be used). The individual specimen that is to be used by the apprentice must actually be captured by him or her. The first bird captured by the apprentice is generally the passage kestrel. Typically,

---

1 William C. Oakes "The Falconer's Apprentice: A guide to Training the Passage Red-tailed Hawk", 2001.

The Red-tailed Hawk

these birds are used only during the period of apprenticeship. However, the truth of the matter is that the red-tailed hawk is ideal for use by both novices and experts, and there are many falconers who never use anything but red-tailed hawks, returning these birds to their natural environment after a period of use. It would be impossible to do this with a bird raised in captivity, since such a specimen would not be accepted by other birds of prey, and would have hunting habits ill-suited to survival in the wild. Another point in favour of this practice is that if, for some reason, these birds that are used for training are accidentally lost, their survival will not be jeopardised since they know how to hunt perfectly well, and they will able to fend for themselves.

In sum, it is required by law in the United States[2], that, prior to being recognised as an apprentice, one must learn by reading and studying a vast amount of information and securing a sponsor to serve as a tutor and mentor during the course of two years. During this time, the apprentice is to assimilate a wealth of information regarding the diet, training, habits, location, and management of the bird that he or she has selected for use during the training period. All of this must happen well before any examination is taken, or before the apprentice takes possession of a bird for personal use. These requirements help assure that every falconer's apprentice received the highest degree of preparation in the proper management of his or her bird.

Taking into account what has just been said regarding the regulation of falconry in the United States, it may be said that the vast majority of falconers have handled a red-tailed hawk at one time or another, whether as novices or experts and masters.

In the United States, red-tails are very popular birds and most falconers have flown and hunted with a red-tail at one stage or another. Although initially, they seem to be the number one choice for beginners there, they can also be seen with master falconers and at many sky trials successfully catching rabbit, hare, squirrel, pheasant, duck and even partridge. There is an extensive amount of literature on red-tails in the United States including several monographic works (see bibliography) though some of the most well-known falconers that have made a name hunting with red-tails are Gary Brewer, Manny Carrasco and Jim Gwiazdzinski.

Gary and Manny live in Texas (USA). They currently hunt together with their respective red-tails. In fact, their friendship itself began as a result of their shared passion for falconry and red-tailed hawks. Gary has written a book about hunting squirrels with red-tails, and he has also produced an introductory video on falconry. Manny has produced a second video as a follow-up to Gary's video, and has also written numerous articles about the practice of falconry with red-tailed hawks.

(The following article first appeared in the December 1998 issue of HAWK CHALK and has been published in this book with kind permission of its author).

---

2 For more information on legal aspects of the United States, please contact the NAFA and your local authorities.

3. The World of the Red-tailed Hawk

Red-tailed hawk with squirrel. Photograph courtesy of Manny Carrasco.

The Red-tailed Hawk

Typical hunting perch of red-tails in the US. Photo by George Robertson.

3. The World of the Red-tailed Hawk

RED-TAIL, WORKHORSE OF MODERN FALCONRY

By Gary L. Brewer, author of Buteos and Bushytails.

If you were going to list the qualities that any falconer would desire in a game hawk, the list would certainly include the following:

- It would be available to most anyone, no matter where they lived.
- It would respond quickly to manning and training.
- It would be aggressive toward game.
- It would have feathers that would seem to be made out of rubber.
- It would fly well in the forest as well as the open country.
- It would be easy to house and transport.
- It would tolerate extremes in temperature with a minimal of consideration.
- It would last for years, improving with age and experience.
- It would respond well in the field and not embarrass you in front of friends and falconers.

This sounds like a tall order, but such a hawk does exist. Sadly, relatively few have truly appreciated her or given her the respect and recognition she has earned. In our country, no other hawk has introduced more people to falconry than she has. I speak of the red-tailed hawk. This article is written as a tribute to this special bird.

The range of various species and subspecies of the red-tail is similar to that of human beings. Virtually anywhere you may (law permitting) with a loaded bal-chatri in your lap is a good place to trap a red-tail and their conspicuous habit of sitting atop the highest perch makes it possible for even the inexperienced trapper to spot them. They are not difficult to trap. A properly baited trap well served will most likely draw a quick and powerful response.

As for the passager, I've never had a freshly trapped red-tail that I could not get to take food on the fist within 48 hours of trapping. Once the ice is broken and provided sensible weight management is practiced, she will soon be looking forward to your time together. She does not require a lot of carrying time to man her down, although it certainly would not impede her progress. The old saying, "...it is not the quantity of time, but rather the quality of that time that matters" applies in training these hawks. A couple of sessions on the creance, and you will suspect that someone has had this hawk before and released it.

With the novice falconer, it is usually the trainer that holds up progress in the training process. Using sound falconry techniques, 2 to 3 weeks is about average to progress from the trap to flying free, assuming weather is favourable.

Competently flown, she is steady in the field an responds to the fist and lure better than most, making it easier to manage her under most any situation you may encounter in the field without any undue loss of training time. If properly introduced, she will fly with a crowd and/or a dog.

Red-tails are selfish and greedy. This may be an undesirable quality in humans, but is the basic ingredient for a good game hawk. The more selfish and greedy they are, the more aggressive they will be toward game. If you put up enough game under them, even an inexperienced red-tail will kill with some consistency, and that consistency will improve with experience. If I had to come up with a disadvantage to flying a red-tail, it would be that they are not able to take feathered quarry in fair flight with any regularity.

There are a lot of true stories out there, and I am sure that on rare occasions a perfect slip is achieved on feathered game but, as a rule, this quarry is not within the effective abilities of the bird, and getting those perfect slips is not within the abilities of most falconers. Mammals such as rabbits, squirrels and hares are well within their ability, and fortunately one or more of these game species is available to most any falconer.

Red-tails in pursuit of game are reckless. They slam into briars, crash into stationary objects and bounce off the ground harder than you think they could survive. They roll and tumble with, and are dragged by, the sometimes-large game they bind to. They do this with no apparent concern or respect for their feathers.

Fortunately, red-tail feathers are so resilient that, unless you fly another species, you will have only rare opportunities to practice your imping skills. In all my years of falconry, I have never actually seen a feather damaged in the course of a hunt. That is not to say that their feathers do not wear. The feathers of a hawk that is fairly flown will wear, but they wear evenly and still give the appearance of being perfect feather. When feather damage or excessive wear does occur, the cause can almost always be connected to the way they are housed, perched or transported.

Red-tails naturally tend to hunt the more open areas for obvious reasons but, being opportunistic, they will also hunt the forest if there is sufficient game to raw them into that habitat. If flown at forest game regularly, they will, in time, develop remarkable muscularity and agility. Forest game such as squirrels, will utilize every obstruction available to save their hide. These twisting, braking and turning flights will work every muscle your hawk has, and she will become graceful beyond belief. I am ever amazed at what an experienced red-tail can do. Frankly, I have seen very few hawks become as graceful as these hawks can be if flown regularly at this three-dimensional forest quarry.

When your facilities and giant hood are constructed thoughtfully, red-tails are kept and transported safely. The main chamber should be spacious and airy. The red-tail will prefer to spend the largest amount of her time during the day in the weathering chamber; therefore it is best to spend the majority of your time and money in the construction of this part

3. The World of the Red-tailed Hawk

of your facility. Your hawk spends most of her time in this facility, and it is my opinion that it can be the most dangerous time she spends if the facility is not constructed with consideration. Believe it or not, it can actually be less expensive to build a safe facility than one that is not. The key is to keep it simple, and with adequate space for her to move about. (For more information for this type of mews – please see the section on Basics of Freelofting in Gary Brewer's website at the end of the book).

The transport box or giant hood could also be simple. My reference for any sized red-tail is a 4 inch x 24 inch x 12 inch box with a padded perch placed 7 inches from the door and 7 inches from the floor. The perch should be mounted such that the bird is facing toward the door. The door should be large enough that she can enter and exit easily.

If the hawk is being flown daily, she can be kept in the giant hood for an extended time. While travelling, I have kept my hawks in the giant hood for two weeks without any problems (of course they should be cleaned regularly). I will also put hawks in the giant hood during very cold nights. Her body heat will keep the interior of the box about 10 degrees warmer than the outside.

Red-tails are biologically tough. They have a temperature tolerance rage of zero degrees (out of wind) to 110 degrees F. making them a good choice regardless of geographical location. They are resistant to maladies, which plague other hawks, making them a good choice regardless of experience level. Their immune system is so effective I would only medicate as a last resort for fear of interfering with it. I have never had a red-tail die of illness, disease or parasite.

Availability, responsiveness, determination in pursuit, variety in quarry, durability, adaptability, disease resistance, hardiness and longevity are all qualities inherent in red-tails. Few other species have this many qualities that make them an ideal game hawk for today's fast lane falconer who has to work falconry into a busy schedule.

Another factor that has prevented the red-tail from receiving her just recognition in American Falconry is that she is one of the hawks that the apprentice is permitted to fly. This has resulted in her being dubbed a "beginner's hawk". Because of this stigma most apprentices can't wait to disguise their experience level by flying something other than the red-tail, Kestrel or Redshoulder. A lot of master and general class falconers would not care to fly them out of fear they might be mistaken for an apprentice.

Again, a true falconer does not judge another based on what he is flying, but rather on what he has been able to accomplish with what he is flying. Falconry is the taking of wild game by using a trained raptor. If game is not being taken, then you are only attempting to practice falconry.

The key to successful falconry is to determine what game is most available and select a hawk that can consistently do the job. Any species of hawk that can handle that job should be held in high esteem, and the falconer who handles her should be respected for his observation and making choices that are obviously based on his commitment to quality falconry.

Regulations limit the apprentice to a choice between the red-tail, Kestrel and Redshoulder, and in most cases the red-tail is chosen. Without the close supervision of a competent falconer who understands red-tails and their quarry, this first experience will likely not be very productive for either hawk or hawker. The new apprentice, regardless of how much he has read on the subject, will be clumsy in the handling of the hawk, disorganized and unsure of himself in the early training process, and he may possibly lack the discipline, determination and commitment to put his hawk in the field as often as he should. Furthermore, he may lack the knowledge and experience to fly his hawk in situations where she has a fair chance to succeed. As a result, the apprentice does not usually set the world of falconry on fire.

During this period of learning, he is constantly reminded to hang in there because, once he moves to the general level, he can fly something else. At the general level, he abandons the red-tail and moves on to something else he may or may not begin having more success with… Often, it's because he now has more experience and is beginning to become a falconer, and does not necessarily have anything to do with the change in species.

Nevertheless, the memory of his red-tail experience is what he will carry on with him into the future. In his mind, he will blame the results of his inexperience and clumsiness on the red-tail he learned on. He will observe other apprentices for the same reasons and this confirms what he already thought. His apprentice picks up on his lack of confidence and looks forward to the day he can get himself a good bird, and the cycle goes on and on.

You cannot pass judgment on value based on species alone. What is a good game hawk? She is an individual hawk of any species that is consistently successful. There are a number of ingredients that must come together simultaneously before a good game hawk can emerge.

The raw material you start with is an individual hawk, of any species, but to succeed, she needs certain desirable mental and physical attributes. Those are the desire to hunt and take prey, some sense of prey behaviour (or the intelligence to learn it), suitable physical characteristics such as body size, flying ability and foot size, and the coordination to utilize all qualities. She needs to be flown at quarry that is plentiful and which is within her capability and to be flown often.

I realize that all birds are not created equal and that goes for any species. I have encountered a few birds that, no matter how well flown, did not show promise of being good game hawks. If she has been properly entered and well flown, it should only take a couple of months to determine if an individual has promise or not. If she doesn't show promise and there is trapping time left, get rid of her and start with

3. The World of the Red-tailed Hawk

a new prospect. Being a falconer does not require you to be a foster parent.

A good red-tail, properly handled and well flown, is as impressive as any hawk. If I had to depend on a hawk to put meat on my table, I would look to the red-tail to fulfil that need.

As a man with a demanding career, as well as with a family who does not understand but is tolerant of my passion for falconry, the red-tail gives me the satisfaction I need and is the least demanding of any species I have flown. In my 20 years of falconry, I have flown red-tails, Harris', Coopers, hybrid falcons and Redshoulder and have close friends who have flown the rest. I can fly anything I want and I do. I wouldn't mind having something a little more exotic as a second hawk to experiment with, but I will keep a good red-tail as a mainstay. She truly is the workhorse of modern falconry.

Remember, the fact that they are so "common" is a testimony to their versatility and ability; I find that very attractive.

God bless and good hawking!

Gary

Cartoon of a typical squirrel hawking day with red-tailed hawks, Gary Brewer and Manny Carrasco. Toon by Manny Carrasco.

Modoc's release – by Charlie Kaiser

Yesterday, March 30, 2003…

I wake at 4 am with a nasty headache. A couple of aspirin, a hot shower, and back to bed… 5 am, I awake again, too hungry to sleep. Why does this always happen on the weekends? Geez… A bowl of granola and back to bed…

I finally awake normally around 7:30. It's a clear sunny day, temps already in the mid 60s, headed for around 80. Another gorgeous day in paradise…

Outside the window, Modoc in her mews is making her dinosaur sounds. Perhaps I'll figure out how to capture the recording I have of it on the computer and post it somewhere. It's a bizarre noise. Anyway, I rise with one thought in mind - It's a beautiful day to release a red-tail…

I've been working towards this moment for a long time. I've known for over a year that it was just about time to let her go. It hasn't been an easy understanding or journey, though.

Modoc was trapped Oct 7th, 1998, in Modoc County, CA. It took over an hour for her to finally get snagged on the trap; she's a wily bird and did everything she could to get to the bait without hitting the trap, but she finally did, even with a 3/4 full crop. Gotta like that pig factor in a hawk…

From the very first time out, she knew exactly what a jackrabbit was - food! She lived to chase those big hares. The only thing she liked better was pheasant. She caught her share of both, too. We had a lot of time together in the field. We developed a bond and an understanding of what each of us wanted out of the situation, and as the years went by, we got better and better at providing for each other's needs.

One of the things I realized this past year was that she was ready for her freedom. She had gotten to where I couldn't fly her with jesses anymore; she'd take them out and eventually learned to leave them on top of a pole rather than dropping them where I could pick them up and put them back in. She also tore through 3 sets of cuffs this season. She really didn't want that stuff on her legs anymore. She had synched into the falconry schema completely and wholeheartedly, though. She hooded beautifully, moved with me in the field, and acted like a well-trained bird even fat during the molt. When I would reclaim her at the end of the molt, I was always certain that I could fly her a couple hundred grams (about 7 oz) heavier than the last year since she was so compliant. She always proved me wrong on that, but it was nice to see that she had comfort with her life.

So, I made an agreement that I would release her at the end of the season. She flew well this year, better than ever. New tactics, higher flight, stronger chasing. She had become a well-rounded hunter by this point, and I was her only limitation. Time, terrain, and property boundaries were what stood between her and pure success. Since those items affected me

3. The World of the Red-tailed Hawk

and not her, it was clear that I was holding her back, and that meant it was time for me to let her go. I had a hard time with that. I know all the reasons that it's the right thing to do - she's a full adult; ready to breed; she's a good hunter, able to survive on her own; she's driven or fought off other red-tails; she knows how to interact with them; she's done her time with me; it's time to let her have her own life back.

All these things and more I know, but while my head said yes, my heart said no. She's been a part of my life for years, and giving that up, knowing I'll never see her again, was extremely difficult…

On March 8, I took her out on one last hunt and she did wonderfully, as usual. I sat with her in the field as she fed on the jack, knowing that this was the last one. I cried a bit; the first of many tears to come. I got her back home and started fattening her up. She seemed to understand what was going on; she put away large quantities of food, and spent much of her days exercising in the mews. I caught her up once or twice and checked her keel; she had cleavage and the muscle on either side of her keel was firm and deep. She was strong, and as my knuckles will attest, her talons razor sharp. She got a bunch of live rats for food to fix rodents as her food image. She learned fast to avoid the biting teeth and kill them quickly.

Saturday, March 29th, 2003…

I know the day is getting closer; I get a date a little more firmly in my conscious, although it doesn't jell completely. I can't bring myself to be that definite; it would mean that the action is irrevocable and I'm not ready for that yet. So I figure, somewhere around the 7th of April; that will be exactly 4 1/2 years. I go to bed Saturday night with that fuzzy concept in mind.

Sunday morning, March 30th.

As I come to full consciousness after the headache and an early breakfast, I realize that no good will come of putting it off any longer. Modoc is ready, the world out there is ready for her, with territories being staked out by resident birds. If I wait much longer, she is going to have to fight hard for a spot. I'd rather make it a little easier for her. The only thing not ready is me. But I'll never be ready. I know that, and come to an acceptance with it. I walk into Pam's office and tell her, "it's a beautiful day to release a red-tail", and I start crying again. (I start crying again as I write this, too…)

Pam gives me a big hug and that knowing grin that says "you figured it out".

I warm up one last quail for Modoc and go out to the mews. She comes to the fist for a chunk of quail, and I put her jesses in for the last time. I take her outside and we burn a couple rolls of film taking the last goodbye pictures while she eats her last free meal. She's going to have to work for the rest of them.

I'm going to release her here at the house, and give her the option to come back if she wants, at least until she gets settled. So Pam takes the bars out of one of the windows; the one where she would perch every morning to watch the sun rise. I take her back into the mews, and cut off the cuffs. She is free. She sits on my fist, giving her bare legs a look. She hops off the fist up to one of the perches and I walk out. I pull up a chair on the patio and wait. I figure she will leave pretty quickly, but once again, she reminds me how little I know; she stays in the mews for over an hour, bouncing back and forth on the other perches, ignoring her favourite which is now no longer fronted by vertical bars. She hits it a couple of times and takes a quick look out, somewhat perplexed by the different look, but goes back to the other perches. Finally, without preamble or even touching the front perch, she flies out the open window, over the house, and up into the liquid amber tree in the front yard. She's free!

Pam and I run out front to watch as she sits in the tree for a while, the scrub jays and bushtits moving closer and closer as they check out this new predator in their midst, their fear minimized by her full crop. She looks around, scanning the skies and the land below. We're on tears of sadness mixed with joy streaming down my face. My heart is broken but buoyed, torn between my loss and her freedom. I know that the pain will pass eventually, but right now it hurts.

As we go back into the back yard, I am amazed at how a mews without a hawk living in it is such a lifeless, empty building, whereas so little time ago it was so alive. We go back into the house and reminisce for a bit, then go out on the deck to sit and read. It's late morning by now, and a beautiful warm day, so we take novels out to the deck and bask in the sun, letting the warmth and light flood into us. After about 45 minutes, we spot a flash out front, and go out to look. Sure enough, there she is, sitting once again in the liquid amber tree in the front yard. My heart cheers seeing her again. She's found her way back, and looks none the worse for wear for her short adventure. She sits in the tree and preens, and we go back to the deck in back, where I can just see her through the tree branches.

After an hour and a half or so, I look up and she's gone again. I check the sky and see her close by, just starting to climb. She does so effortlessly again, and I watch with binoculars as she climbs the staircase of rising air into the sky. She soars off to the north, and I watch as she continues to circle, climb, and head downwind. She finally specs out in my 10x Nikons, and I drop them from my eyes as she disappears. I think to myself, I may never see her again.

We spend the rest of the afternoon sitting around reading, taking it easy. Of course, I'm checking the tree and the sky every half hour or so, but she never returns.

She is truly gone, and a part of my heart has left with her. I hope I will see her again someday, but I have a feeling I won't.

Many of you know of Modoc, you know what she looks like. Some of you have hunted with her. You know how I feel about her. If you see a rufous-morph red-tail with a short 3rd primary on her right wing flying around your area, say hi to her and tell her I love her….

**Vaya con Dios,
Modoc…**

3. The World of the Red-tailed Hawk

A nice close-up of Modoc. Photo by Charlie Kaiser.

Puerto Rico

Even though Puerto Rico is part of the United States, the legal status of falconry on this island is at present in abeyance, with the Puerto Rican legislature having granted the sport legal approval in 2005 but approval, by the nation's Department of National Resources, is needed before the sport can achieve legal status in the island, and this has not happened yet. This was the situation in 2004, when I first published this book, and as far as I have been able to confirm, I understand that the practice of falconry in Puerto Rico is currently still illegal.

It is in Puerto Rico where the falconry master Juan Manuel Iglesias has lived for over 20 years, and where he is the director of a centre devoted to the study and preservation of birds of prey. He has conducted research on more than 100 individual specimens which have been brought to his centre in order to be rehabilitated. For the past ten years, Juan Manuel has been the only person in Puerto Rico who has had federal permission to rehabilitate birds of prey. He utilises the techniques of falconry in order to rehabilitate birds of prey, eventually returning them to their natural environments.

Juan Manuel Iglesias has identified the following subspecies red-tailed hawk in Puerto Rico: B.j. borealis, B.j. umbrinus, B.j. Fuertesi and, finally, the fourth subspecies of red-tail discovered by him, and which is native to the island: B.j. portorincesis, which is described in the second chapter of this book. Iglesias is a man who possesses a vast knowledge of red-tailed hawks[3].

There are currently not many experienced falconers in Puerto Rico. Those falconers that do reside on the island may be divided into two groups. One group of falconers is called Falcon Sur de Puerto Rico ("Puerto Rican Southern Falcon") and is located in the south of the nation. Its director, Mr. Laredo González, was born in Cuba. The other group is called the Puerto Rico Hawking Club, which is located in the north of the island, and the director of this latter group is none other than Juan Manuel Iglesias himself. The total number of falconers on the island, including the members of these groups as well as amateur falconers, likely amounts to no more than 20 persons.

The animals that may be legally hunted in Puerto Rico are generally limited to birds, including those of the aquatic variety such as ducks, which may be hunted during a designated hunting season. In addition, red-tails may also feed on woodcocks, snipes, turtledoves, and other types of pigeons, in regions where this is allowed. The mongoose is one of the few mammals that one is allowed to hunt throughout the entire year.

Juan Manuel Iglesias, who's photographs of the B.j. Fuertesi Roka appear

---

3 The variants of the subspecies of Buteo jamaicensis and the newly discovered subspecies in the island of Puerto Rico have not yet been officially classified or acknowledged by the A.O.U and are the result of many years of research in the island by Juan Manuel Iglesias who has given me permission to publish a few highlights for this book

3. The World of the Red-tailed Hawk

on pages 121 and 122, set up a trial in which this bird hunted a dove that was initially released some 100 metres (109yrd) away. Roka captured the pigeon at a point 200 metres (219yrd) from its own original starting point (i.e., the bird was able to cover 200 metres in the time that the dove covered 200 metres or 219yrd). In this trial, Roka began its pursuit from atop a wooden telephone pole that was about eight metres (26ft) high.

In another pursuit of a dove, Roka covered one full kilometre (0.6 mi) during its pursuit. On another occasion, it was initially stymied after pursuing a dove for 300 metres (328yrd). After hovering near the area where the dove had obtained temporary refuge, Roka once again spotted its quarry and finally chased it down after traversing another 600 metres (656yrd). Roka has attained speeds typical of an accipiter while pursuing mongooses. In addition, this extraordinary specimen of red-tail has chased down turtledoves, teals, and other birds.

Juan Manuel has also trained a specimen of the subspecies Borealis, which he named Bony, and in one of his slips, he

*Here we can see Roka, female eyass variant of fuertesi subspecies with a mongoose. Juan Manuel has already caught more than ten mongooses, although this type of hunt is quite complicated. They are more difficult than rabbits as they can defend themselves more. They move in short distances amongst the shadows and at times, you may bump into them with your feet yet not really see them and realise much too late, when the hawk has already bated off the fist. They are very aggressive and put up a good fight. If the hawk doesn't grip them well with her feet, they can manage to break free and attack her.
Photo courtesy of J. M. Iglesias.*

The Red-tailed Hawk

Roka and a guinea-fowl. Photo courtesy of J. M. Iglesias.

Another photo of Roka, female red-tailed hawk, variant of the "Fuertesi" subspecies in Puerto Rico, seen here having caught a large rabbit, similar in size to a hare. Photo by J. M. Iglesias.

3. The World of the Red-tailed Hawk

Female red-tail, borealis variant. Photo by J. M. Iglesias.

chased down a dove at ground level for some 20 metres (22yrd), after which the dove cut into a small forest of mangrove trees, and Bony immediately soared over the trees and was forced to penetrate into the forest. After failing to return when called, José Manuel went into the forest and found the red-tail depluming the dove that was tightly grasped between his talons.

In another slip, Bony pursued a Muscovy duck (Carina moschata) that came out from a narrow river with a very strong and powerful ascending flight, at about 10 metres (33ft) away and binded on to it instantly in the air. These ducks are native to tropical American (Mexico and Argentina) and they are very similar to geese. Mostly, these ducks tend to escape from farms and interbreed with wild species.

During another outing, with a tiercel named Champ, at times he would catch moorhens in flights of only 50 to 100 metres (55–109yrd), and he also managed to pursue them bolting from my fist at distances of 200–250 metres (219–273yrd). He caught them with such ease and in so many different ways, that even difficult slips proved to be successful and always surprised José Manuel.

However, Juan Manuel was not particularly successful in training specimens of the subspecies umbrinus, although these birds did display a high degree of aggressiveness, as well as considerable courage, in pursuing large prey.

The Red-tailed Hawk

This is a photo one of the few Canadian falconers who uses the red-tailed hawk to hunt. Bob Smurfitt is seen here with his passage red-tail which is widely found in Canada. Bob is holding a pheasant captured during a day of hunting in Alberta, Canada. Photograph courtesy of Roy Priest.

3. The World of the Red-tailed Hawk

Nest and typical habitat of red-tailed hawks in Canada. Photograph courtesy of Roy Priest.

Canada

The red-tailed hawk has not been utilized by Canadian falconers nearly to the extent that it has by those who practice the sport in the US. The practice of falconry itself is rather more limited in Canada than it is in the US, although the country does have three fairly large falconry societies: The British Columbia Falconry Association, The Alberta Falconry Association, and the Saskatchewan Falconry Association.

The fact that Canada does not have the same system of apprenticeship as the US has probably led to the red-tailed hawk being less used by falconers in that country. However, red-tails of diverse subspecies are rather common in this country, and may be observed, as in the US, on telephone poles alongside highways or in rural areas. In British Columbia, red-tails of the harlani subspecies may be found in the northern part of the province. Kriderii may be found in the prairie provinces and borealis red-tails are present throughout the rest of the country.

Roy Priest lives in Lardner, British Columbia, and began practicing falconry in 1963 after attending a falconry exhibition by a local falconer. He joined the British Columbia Falconry Association and began going hunting with his first bird, a red-tailed hawk aged 13 years. This was the first bird that he trained and it remains, in his opinion, the most gentle and noble bird that he has ever handled. Since that time, he has also flown kestrels, peregrine falcons, prairie falcons, merlins, lugger falcons, cooper hawks, Harris hawks, and goshawks, but he continues to have a special place in his heart for the red-tailed hawk.

Europe

Buteo jamaicensis continues to be relatively unknown in large parts of the European continent, although it has gained certain popularity in the United Kingdom in recent years, where various red-tail subspecies may be found, and where red-tails are now both raised in captivity and very much utilised in falconry.

United Kingdom

The use of the red-tailed hawk—although not nearly as common as in North America—is quite widespread in the United Kingdom. During the past twenty years, red-tails have become one of the most popular birds chosen by falconers. One of the reasons that this is the case is that red-tails are aggressive when hunting and yet gentle with their handlers—as long as they have been properly trained: this is a very important quality for novice falconers. This hawk can also be used for both high and low flying, has a more agreeable temperament than the goshawk, and hunts with a similar level of dedication. Red-tails work well with both beginners and experts, and are not likely to become boring for their handlers.

It must be said that, like the Harris Hawk, which is also recommended for beginners because of its quiet and gentle character and ease of handling, the red-tailed hawk initially appears to be a little "monster". However, after several days, it usually becomes the most tranquil and noble bird of prey that one could possibly imagine—much more so than the Harris Hawk—while displaying remarkable quickness, agility, and flying skills. Anyone who has ever flown a red-tail will understand very well what I am talking about.

In the United Kingdom, red-tailed hawks are used to hunt rabbits, pheasants, hares, and even geese—just as in the United States. The most common subspecies in the UK are both the Buteo jamaicensis calurus and the Buteo jamaicensis borealis with some degree of intergrades also available from imported stock. The fuertesi and the harlani are also found on the island, but in far smaller numbers. In recent years, I have also seen a few "ferrutails" or hybrids of red-tail x ferruginous buzzard here as we shall see later on in this chapter and have been very impressed. When I heard from falconer and friend Ben Long that he was going to be training a large female ferrutail this season, Gwen, I just had to go over there and see for myself how she was doing throughout various stages of manning and training. I was very lucky to be able to see this as there are not many ferrutails around and I had been quite curious for a while. I was in fact quite anxious to see what she would look like and to meet her. In sum, she was just a larger red-tail with a bigger head, bigger feet than most ferruginous hawks. Quite sweet and noble mannered, yet powerful and also a little intimidating perhaps for some. She really reminded me though of "Bandit", the large eastern tiercel red-tail that I flew and trained a few years ago and left me once again with that feeling… these birds are really just something else!

3. The World of the Red-tailed Hawk

The author, amazed at how well behaved Gwen was after having arrived only a couple of days before. Easy to deal with, extremely gentle and quite calm and relaxed sitting on anyone's fist, she had responded very positively both in manning but was not the biggest fan of cameras. Photo by Arjen Hartman.

Going back to red-tails, in the UK, these are generally used to hunt from trees, as are Harris hawks. However they are also used for other types of hunting, since different types of terrain and different situations require different techniques. Above all, it will be the terrain where the hunt takes place that determines the type of hunting technique that is most fitting. Depending on the situation, posts, perches, or poles are used. The red-tail sometimes begins pursuit of its prey from an initial perch on the fist of the falconer himself. At other times, it will fly closely behind the falconer. In the UK, the use of ferrets along with red-tails is very common in hunting rabbits and hares. At times, dogs have also been used with red-tails, an arrangement that seems to have worked quite well (while the use of dogs with Harris hawks has often had unfortunate results, since these birds tend to confuse dogs with their principle enemy; coyotes) while it is unlikely that a red-tail maintained at proper weight will attack a hunting dog. Red-tails, may, however, attack ferrets, at times and confuse them with furred quarry. If a decision is made to fly in a forest area, the standard procedures is to launch the red-tail toward the trees, and to walk beneath them along with the hunting dogs, flushing the prey towards, the bird as it follows, and thus allowing the possibility that it can totally surprise us and dart toward an animal that the falconer has not seen.

Red-tail Experiences – Ben Long

My first taste of flying the red-tail was around the mid-1970s. I had been flying shortwings, gosses and spars, for some years with reasonable success, and was asked by an acquaintance to train a female red-tail. I was aware that accipiters were supposed to be harder to deal with than buteos, but I quickly realised that, although generally easier to train and maintain, red-tails were certainly different.

A spar or gos shows nervousness by bating; constantly, if you are unwise enough to let it. A red-tail, however, will often just sit there, apparently unconcerned. In fact, they are just too scared to move.

It soon became apparent that I needed to reduce the red-tail's weight by a considerable amount, and that they can maintain a fat weight on a very small amount of food. Due to their slower metabolism, it would be reasonable to suggest that a red-tail can sustain their weight on virtually the same diet that would keep a spar level, although the red-tail has a body mass some five or six times that of a spar. The weight was gradually reduced, and she obviously became increasingly keen to jump to the fist for food. I had reduced her weight by some 6oz (say 170g) and she was jumping up to some 10 metres (33ft), on a creance, to the glove in the garden and mews area. In fact, she became almost manic to get to a morsel of meat, gripping and tearing at the glove to gobble down the food. At this point, it seemed pertinent to start calling-off lessons on a longer creance, so I started taking her out of the garden to a field no more than 50 metres (55yrd) away form her territory. The first time I tried this, I rolled out 10 metres (33ft) of line from an appropriate field gate, tied her on, placed her on the gate and literally ran to the other end of the creance, expecting at any second to feel her claws on the back of my head. As I turned, the meat was in the glove, the glove shown, and I saw that she had her back to me, totally uninterested in food. How could this be? Over the next few days I repeated the same exercise, and was baffled by similar results. It seems obvious with hindsight, but of course she was just scared by the new surroundings and showed that typical red-tail behaviour of just sitting there. It needed more manning and even more weight off to achieve the desired distance and response, and then she was ready to fly free, whereas my experiences with accipiters showed that, if the weight reduction and manning was sufficient for them to jump to the glove immediately, they would do the same just about anywhere.

Another thing to bear in mind when comparing the broadwing and the shortwing, just as it says, is the difference in conformation between the two families. Accipiters have a wing shape which, combined with their longer tail, enables them to chase their quarry in a short dash, twisting and turning to follow their quarry. The buteo is just not designed that way. Its wings are long and wide for soaring and

gliding effortlessly in thermals or any upward currents of air, but level flight is relatively slow, and catching quarry relies on stooping from a height to achieve speed.

As we know, all calling-off lessons should be carried out by the hawk being called to the fist into wind, to keep her ground speed slower, which is her speed relative to the stationary glove. Flying speed is that relative to the air she is flying in, which is affected by the wind. In simple terms, if the hawk is called to the fist with the wind behind her, her flying speed (say, 50 km/h) is added to the wind speed (say 20km/h), the hawk will be approaching the fist at 70 km/h, which could result in her losing confidence, missing the glove and ending up hitting the end of the creance or landing in a tree some way off. Conversely, the hawk flying into wind will be closing on the glove at flying speed minus wind speed, so in this case 30 km/h, which is much more manageable for the inexperienced hawk.

With the first red-tail, as her flying distances increased, I noticed that if the wind was quite strong, she struggled to make it to the glove at all. I began to realise that the red-tail's flying speed in level flight is quite poor, and indeed if the wind speed was similar to her speed of flight, she could even appear to be flying backwards!

I found no problem in introducing the red-tail to catching rabbits, firstly by pulling a dead one along on a string. When it came to flights at quarry though, even a slight headwind was enough to prevent her from catching a rabbit on level ground, and even with no appreciable wind, a flight up a slight hill had the same effect. So, they have the willingness to catch quarry, but not the same ability as a shortwing, at least when it comes to a chase.

Height and wind direction are the key to successful hunting with a red-tail. It would seem obvious, therefore, that flights from trees would be the most likely to succeed. However, success should not be measured by the amount of quarry caught. Instead, a good chase, whether or not it results in taking quarry, is far preferable to any amount of prey caught by the hawk "flopping" from a tree, where there is no chase, and the quarry is usually not even seen by the spectators until the hawk is on it.

After some trial and error, I found that with this and later red-tails that I flew, they were always ready to do their best to chase and catch prey. If they chased and missed, it was quite likely that they would swing up into a tree to take a better look at proceedings, and would easily spot prey from their high vantage point. They would sail off sometimes in a shallow glide and attempt to catch the smallest of prey, which they could easily see at a surprising long distance. That is not the most satisfactory sort of hunting, and I have seen other falconers' red-tails totally spoilt by allowing this form of flight. Red-tails, in common with most hunting raptors, will naturally take the easy option, and chasing prey from the fist is unnatural for them anyway. It is not

The Red-tailed Hawk

unknown to find a red-tail which will sit all day in one or other tree, waiting for quarry to come along, while the falconer tries all forms of enticement to get the uninterested hawk down. So, don't let them do it. It needs some discipline, but <u>always</u> make them chase the quarry from the fist. <u>Always</u> call the red-tail down to the fist immediately after an unsuccessful chase. <u>Never</u> be tempted to leave them there, just in case fresh quarry is found. When they do catch something, <u>always</u> reward them immediately with a large crop of warm meat, even if you have only just started hawking that day. By the way, a common trait of the red-tail is that they are never keen to be taken off the quarry they have caught, and will grip onto it as if life depended on it. Be very careful at this point, for if they think you are going to take the food away, they are likely to grab your bare hand, and you will <u>not</u> get free without assistance, some blood loss and a good deal of screaming.

Many falconers say that red-tails are slow and lazy. They are certainly not lazy, but they are not the fastest thing on wings. So, what's the answer?

Keep the red-tail on the glove, and take the glove to the highest point, of course. Use the advantage of walking on hills and banks while your dog or helpers flush quarry below. Always remember the wind - it can be the red-tail's ally. Keep it to your back as much as possible, recalling that in this way flying speed + wind speed = ground speed. A running quarry is relatively easy for the pursuing hawk in this way. Even a pheasant, flushed from below and downwind, is a possible achievement, as the tailwind will favour the larger wings of the buteo.

Last, but certainly not least, use their ability to soar - it is, after all, what they are best at. As we said, the buteo can stay aloft on the slightest upward air current. As the wind hits the slope of a hill which faces it, the air mass is forced to go over the hill, and in this rising air the red-tail can soar with no effort. It is a little hard to introduce the inexperienced hawk to it, but reward is the key. Cast her off into the wind, and run down the slope, thus encouraging a soaring flight. Within just a few seconds, bring her in to the glove with a small amount of meat on it. Repeat this several times, with longer intervals each time, but never so long as to allow her to turn away and land somewhere, or the day's lesson is wasted. Soon, when maybe half a minute's soaring is attained, a dragged dead rabbit can be used, which should be pulled by an assistant from further down the hill. This will induce a long gliding stoop, and a moderate cropful of meat should be the reward. Very much like flying falcons at game, the red-tail soon recognises that a dog or beater is the source of quarry, and will soon be soaring over them in expectation of prey being flushed. Success breeds success, and the time they will stay airborne will get longer and longer.

Don't compare the red-tail with other falconry birds. They have some disadvantages, but use their attributes to find success with this unique raptor.

And be careful of those feet!

3. The World of the Red-tailed Hawk

Ben Long is a well-known British falconer and artisan who has also flown red-tails in the United Kingdom. He became interested in falconry at a very early age and in 1971, with just sixteen years of age, bought a common kestrel which would be his first ever falconry bird. Ben became more and more involved in falconry, becoming a member of the British Falconer's Club and was even making falconry equipment professionally by 1973. Since then, he has built a solid reputation for his experience not only as one of the world's finest falconry equipment artisans but also as a falconer. Ben has flown and trained countless birds of prey, from the common buzzards, red-tails and Harris hawks with which he taught falconry courses to beginners, to falcons (including hybrids), accipiters and various other birds of prey. However, Ben has confessed that his favourite birds of prey are peregrines, sparrowhawks and red-tailed hawks.

As you will see in the following article, he thinks highly not only of red-tails, but of red-tail hybrids such as the ferrutail. I met up with Ben a couple of times during the 2007 hunting season, when he flew a marvellous female ferrutail named Gwen and was once again, amazed and captivated by these birds. I think with hybrids, if one is lucky, one can either get the very best of two birds, or indeed, as I have seen at times, the worst out of them. I am happy to say (and quite jealous may I add!) that the ferrutail is definitely one combination I shall be watching out for in the future as like red-tails, they also have a lot of potential awaiting to be discovered.

Fun with the Ferrutail – Ben Long

The red-tail/ferruginous hybrid has only been produced by a handful of UK breeders. Until this year, when I was given the one I have now, a female which became known as "Gwen", I had never seen one. She was bred by my friend Anthony Haley in Scotland, and was the result of a natural breeding between a female red-tail and a male ferruginous. He bred them for the first time last year, and was interested in my flying one to assess their capabilities.

At first sight Gwen was pretty impressive, being very much the colour of a young red-tail, but somewhat larger. Similar to a ferruginous, her head is a few sizes larger than a red-tail, even though a red-tail's head couldn't exactly be said to be small. To give an idea, I also have a couple of gyr/saker hybrids flying at 2lb 11oz (1,219g) or so, a very similar weight to a female red-tail. Their hood size is 104, whereas a red-tail of the same weight would take a hood sized around 110. Gwen's hood size, although she is only a few ounces (50g) heavier, is 120, which is exactly halfway in size between the red-tail and a typical male golden eagle, which would fly at about 7½ lbs (3.401g)! Of course, her mouth is also correspondingly large, and leg feathers continue about halfway down her tarsus, instead of

The Red-tailed Hawk

finishing at the hock like a red-tail. Luckily, she takes after a red-tail in her foot size, as those of a ferruginous are notably (some might say ridiculously) small, compared to their body size.

A nice close up of Gwen. It does in fact, just look like a great eastern red-tail, except of course for the huge gape. Photo by Arjen Hartman.

As a hood maker, I am often quoted "fat" weights of newly acquired hawks, in order to gauge the hood size. These weights are frequently impossibly high, as the hawks have been weighed straight out of the pen, when they could have eaten a large meal shortly before weighing, and probably with a bit added on for luck. Particularly with larger hawks, it is much better to weigh them at least 24, and preferably 48 hours, after taking up the bird in order to get a genuine weight to work from. At first weighing, two days after she had last been fed, Gwen was 3lb 4oz (1,473g) and I anticipated a hard slog over some weeks to reduce her weight sufficiently to make her responsive. When I first took on the fist, although alarmed, she was pretty steady and disinclined to bate, and luckily continued to be so during training. At first, I decided to attempt to hood her, but she took an instant dislike to it, although probably more to my hand than the hood. I desisted with the hooding after a couple of days, and on reflection feel it would have been better to have got her very used to the touch of a hand over the training period, then introduced the hood afterwards. It isn't the usual way, but could have been better idea.

She took about three days and some ingenuity to coax her into feeding on the fist for the first time. She eventually decided it was OK to bite and hold the food in my hand, but took a very long time to pluck up the courage to take a piece out of it. Once she did, progress was considerably faster. As she was so steady, manning was a fairly rudimentary affair, entailing not more than 15 minute's carriage per day. I was able to put her to weather on a bow perch after only a few days, although as usual it took a few more days to work out that the bow perch was for sitting on. It was relatively painless to pick her up, even though it took quite a while before she would jump readily to the fist.

It was, of course, quite hard to reduce her weight, and her daily ration had to be only about ½ ounce to make some progress. Once she got below 3lbs (1,360g), she began to step up to the fist, and once she was down to 2lb 12oz (1,247g) she was coming to the fist almost immediately at about 40 yards (37m). I felt this was quite a moderate

3. The World of the Red-tailed Hawk

Ben Long and Gwen ready for a training session after a nice sunny day in the garden. Gwen gave no trouble at all being picked up or going back to her perch, having arrived only a few days before and could already be picked up by anyone with a glove.
Photo by Arjen Hartman.

Gwen flying well as soon as Ben called her to the fist. Photo by Arjen Hartman.

The Red-tailed Hawk

A nice shot of Gwen flying very low to the ground. Photo by Arjen Hartman.

Gwen flying well as soon as Ben called he Gwen taking the bunny or dragged lure. Photo by Arjen Hartman.

3. The World of the Red-tailed Hawk

amount to have to take off, for such a large bird. By this time, she would also attack a rabbit lure quite readily, so I flew her loose. This had taken only 3½ weeks from first starting the training, and I think if I had been a little more ruthless with the weight reduction she may have been flying free in less than 3 weeks. Once she was flying, I began to ease her weight up, and within about 10 days she was still responsive at very close to 3lbs (1,360g). She also flew happily to anyone with a glove, even total strangers. She has not so far been free with her feet, although I am sure that in time she will become more possessive with food and will need to be handled more carefully when food or quarry is around.

An attempt was made to train Gwen to the kite, which I hoped would make it easier to get her to soar, and to improve her fitness. She cottoned on very quickly to grabbing a winged lure suspended from my hand, so when I attached the lure to the kite string she had no trouble in taking it in a single straight flight up to a height of 20 feet (6m) or so. Unfortunately, she didn't get the hang of circling to gain height, so I abandoned to kite in favour of getting her to take quarry at the earliest opportunity.

Entering Gwen to quarry proved a little more difficult, although not due to any lack of aggression on her part. As soon as she was taken to a place other than her training fields, she was "on tiptoe" waiting for something exciting to happen. Initially, it was hard to find short flights due to the abundance of cover and the quarry being reluctant to disclose

Once again, a nice shot of Gwen coming in to the fist. The author was amazed of how much she resembles her own eastern male red-tail. Photo by Arjen Hartman.

135

themselves. She was taking on long slips in the fashion of a ferruginous, low to the ground and surprisingly fast, but the quarry was dodging into cover just before she arrived. These were mostly rabbits, although her very first slip was at a flock of rooks about 150 yards (137m) away. Interestingly, she stood looking intently at them for a while, and as soon as a fight broke out between two rooks she instinctively realised they were temporarily at a slight disadvantage, and only then did she take on the slip. Of course, they were all up and away when she was little more than halfway there, but she carried on to sit up in the next hedge, and after only a short pause returned 200 yards (183m) or so back to the fist.

Lamping was the answer. I personally don't care for the practice very much, but with Gwen it worked well to give her the initial confidence that she needed, and in a couple of outings she had caught three rabbits, and was fearlessly slamming into heavy cover after her quarry in the manner of the best red-tails. I had moved her weight gradually up to 3lb 2oz (1,416g) as her muscle mass increased, and she was still flying aggressively.

So, that is the story so far. Gwen is still, a the time of writing, only six months old, and I would expect her to go on developing for the next couple of seasons. As she gets older, she may well turn a bit more "grumpy" as red-tails and other broadwings tend to, but we shall see. She has now taken well into double figures of rabbits, and has also taken a pheasant, although that was something of a surprise attack. I would hope that she will start to take on proper flights at pheasants, which can be successful if there is an appreciable wind up the tail of the pursuer. It certainly works that way with red-tails, where the wind in their big wings greatly assists them, and I would say that the ferrutail has the edge over the red-tail when it comes to flying speed. Hopefully, it's also possible to get Gwen to soar in due course, which is their natural mode of hunting.

We can expect her to have a life-span of up to 30 years, so it's an ongoing story.

France

When I first published this book, back in 2004, our French neighbours seemed to be completely oblivious to the existence of the red-tailed hawk and had never even heard of it. I have been in touch again now, a few years later, with some French falconers and apparently now there seem to be one or two red-tails used for displays, but that's about it.

Germany

In Germany, falconry is heavily legislated. Aspiring falconers must first obtain a normal hunting license through a hunting examination before they can apply for the falconry test. However, not many are granted falconry licenses as such; the test is so demanding that, on average, every second person fails. Those who do pass the test must also obtain permission to hunt in a given district or hunting ground. Germany also only allows for three of the 15 native German raptors to be used in falconry: the golden eagle, peregrine and goshawk. In addition, falconers are not allowed to own more than 2 individuals of these native species at a given time. According to official figures given by the IAF, the most commonly flown hawks in Germany are Goshawks (about 60%) and then peregrines (about 15%). Red-tails (an important number are also captive bred here and exported for falconry abroad), together with Harris hawks, eagles and falcons make up the rest.

The Netherlands

Although the Netherlands was the centre of European falconry for centuries, currently it has very strict limitations on falconry and the number of falconers actually allowed to hunt. However, although this situation makes it a little difficult to practice the sport, and creates problems amongst falconers, it nevertheless still continues to be extremely popular. One can find much material on falconry in the Netherlands, from collections in museums such as the Valkenswaard museum, to books, and many websites, falconry displays and a large number of breeding centres. Five clubs exist, the largest two being the Nederlands Valkeniersverbond Adriaan Mollen and the Valkerij Equipage Jacoba van Beieran.

Falconry was originally regulated in the Netherlands by the Vogelwet law of 1936. On April 1st 2002, the Flora en Fauna law came into force and continued with the same principles, developing this law further. Basically, the Netherlands grants a limited number of falconry licenses known as "valkeniers akte". Since the 1980s, the Ministerie van LNV established the official number of licenses granted to falconers at 200 and falconers are only permitted to fly goshawks and peregrines at quarry. There is a waiting list and licenses have been granted at the rate of only 1 or 2 per year. In order to get on this waiting list and be considered for a falconry licence, you must spend two years with a mentor from an acknowledged falconry club or organization and you must have also completed a theoretical falconry course. This is still no guarantee and many falconers remain on the waiting list for years.

As the current legislation makes it almost impossible to hunt and practice falconry as such in the Netherlands, there are really only 2 possibilities open to most falconers: either keep a bird only for training and hunting with the lure (as well as just for keeping or breeding) or like Denmark, go and hunt abroad in the surrounding countries. As a result, many breeding centres have blossomed in the Netherlands and there are many falconers also offering falconry displays, although this is also highly regulated by the Flora en Fauna law through article 12 with a limitation on the free-keeping of birds of prey in order to protect (and not scare away) wild species of birds. Sadly, therefore falconers cannot hunt with red-tails in the Netherlands and though many falconers do own and keep red-tails, these are almost always used in displays or for breeding purposes.

Denmark

Sadly, falconry has been banned in the country since 1967, when the Danish Game and Hunting Act was revised. It was then decided that falconry no longer should be allowed in Denmark, though for some years a few falconers were granted an exemption. The final ban on falconry came May 6, 1993 when the new Game and Hunting Act came into force. Previous to the reading in The Danish Parliament an emotional and subjective debate had taken place. Falconry was not the only issue. There was a desire to deprive the falconers of their birds.

Morten Clausen, one of the founders of the Danish Hawking Club, explained to me a few years ago that despite efforts of the club during the last few years and after speaking with several Danish falconers this summer at the International Festival of Falconry (UK), regrettably the situation has not changed. Falconers are still allowed though to keep their birds and train them or breed them (and there are many red-tail breeders also in Denmark) but if they wish to hunt with them, they must do so out of the country as in Denmark it is illegal. Falconers wishing to hunt or practice falconry as such are forced to travel abroad to England, Scotland or Germany for example. There are however, many red-tail fans in this country.

Italy

Although falconry almost died out in Italy in the 1900s, shortly after it experienced a revival due mainly to the publication of literature on the subject and interest has been growing ever since. Currently, Italian falconers fly longwings at pheasant, partridge, quail, crows and magpies, and goshawks at rabbits and hares. Like in most Mediterranean countries, it is very difficult to practice classical game hawking, due to competition for land with strong shooting interests. However, just like Spain, there is a lot of interest in falconry and birds of prey and there are 31 official falconry clubs in Italy that are affiliated to one of the three large falconry federations or unions.

Falconry is regulated through the law 157/92 "sulla caccia col falco e il suo allenamento". Article 23 states that the use of captive bred birds prey belonging to native species (these mainly being peregrine falcons, merlins, goshawks and sparrowhawks) is allowed. However, this law also states in clause 3 that the training of these birds of prey during closed hunting seasons can be allowed prior request of permission from the provinces as long as they do not in any way capture wild prey. In open hunting season, falconers must follow the same rules as gun hunters. In addition, falconers will need to have the relevant documentation for their birds (CITES certificates) and pay for a third party liability insurance.

Basically to be able to fly a bird of prey in falconry in Italy, falconers must be over 18 years of age and need to obtain a hunting license "licenza di caccia". For this, those who aspire to be falconers will need to take an exam in order to hunt with those birds that were described in the previous paragraph. Among the content areas covered in this exam are the following: hunting legislation, zoology for hunters, the protection of the natural environment, principles for safeguarding agricultural production, hunting weapons and ammunition, hunting rules and conduct, and first aid.

The exam tends to be more difficult in northern Italy, and a little easier in central and southern parts of the country. It contains both written and oral sections, and includes the assembly and dismantling of a hunting weapon. If the exam is passed, the corresponding province will send a letter of permission to the hunting authority in the province. This letter may form the basis of a request for a hunting license from the police department in the zone where one is seeking to hunt. In addition, many regions also require for the aspiring falconer to perform a shooting test even if they never intend on firing a gun.

All the rest of the regulations are very similar to the Spanish requirements and falconry is regulated by regions or "provinces". Thus, birds used in falconry must bear a ring on their ankles that indicates that they have been raised in captivity. This ring is used in association used for the purpose of registering and identifying birds of prey—normally, this is the CITES, if the bird requires it (red-tails and Harris hawks do not require the CITES). In this document, all pertinent information regarding the bird may be found. One must always have a photocopy of this document on hand when transporting a bird used in falconry. In addition to the CITES, there is, as in Spain, the document of cession. All previous cession documents must be attached to the most recent document of cession. The acquisition of any bird of prey must be communicated to the provincial department within 15 days of its occurrence.

With regards to red-tailed hawks, since they are not "native" birds, they are not legally allowed to be used for falconry as such or "hunting" but like in the Netherlands or Denmark for example, can be kept and flown for training. Most red-tails in Italy are actually kept by biologists for research purposes or by breeders, but legally, you cannot hunt with them.

Among the relatively few Italian falconers who would prefer to use red-tailed hawks is Alessandro Morasutti, the dean of falconers in the region of Friuli. He is a member of the Falconry Association of Friuli-Venezia Giulia, and he came to my attention thanks to Marco Calistri. Morasutti flew a female red-tail that he later gave to a friend. After the bird had not seen Morasutti for a year, she recognized his voice when they were once again near each other, and called out to him continuously in the manner characteristic of red-tailed hawks until she was able to see him.

According to Morasutti :
"Its brown eyes reveal a high level of aggression, but its expression

The Red-tailed Hawk

becomes gentle when it looks at its owner. These birds are relatively easy to train, and have shown themselves to be adaptable to the demands of the falconer, while demonstrating a constant affection for him or her. This particular female flew at a weight of 1.3–1.4 kg. (45.8–49.3 oz), and at times was even heavier. Yet she was a killing machine, although it is possible that she was not aware of just how powerful she was. I say this because, at times, instead of ferociously thrusting herself toward her prey from my fist, like a goshawk would she briefly became highly excited, during which time she seemed to be taking into account her abilities and making the decision to attack, and hiding herself: all of this made capture more difficult. This kind of phenomenon is especially observed during the initial sprint when the prey is a pheasant, with the red-tail lengthening its thrust toward its prey until the capture is really no longer a "surprise". Such a technique is hardly in accordance with the aesthetic canons of falconry".

The red-tail shows itself to be a good deal more versatile when pursuing rabbits, hares, and cottontails—or other wild animals that are not particularly large. In these cases, the attack of the red-tail is more immediate, the capture more secure, and the style employed both elegant and athletic. It should be borne in mind that this buteo, in contrast to the goshawk, prefers wide open spaces: this allows it to carry out long pursuits. At the moment of capture, the red-tail reveals the full measure of its aggressiveness, transforming itself into a miniature eagle. Its tender side is, at such moments, nowhere to be seen, as it fiercely defends its prey, and will

Baby red-tail harlani, one of the first bred in Italy. Photo by Andrea Brusa.

3. The World of the Red-tailed Hawk

attack any animal that attempts to seize it. When the battle is over, the red tail, straightens its feathers and its posture, and once again appears serene, majestic, and passionate.

Andrea Brusa, who is widely respected in Italy for his experience with Harris hawks, and who is the founder of the Harris Club and the current president of the Yarak Falconry Club, has had the good fortune to begin raising the very

Here we can see Fede, Andrea Brusa's son, who was not even 3 years old with one of the harlani red-tails. As we can see, a properly bred and trained red-tail is definitely not a danger if one is responsible and not at all a "killing machine" as many falconers state, having only dealt with imprints. Photo by Andrea Brusa.

first Harlan's hawk in captivity within Italy. The photo of this bird can be seen below. The parents of this chick were both born in Canada in 1992, and were both previously used in falconry. This chick was first raised by an incubator, and then by Mr. Brusa until the tenth day following hatching. After this, it was raised by a male and female Harris hawk.

Spain

The Iberian Peninsula counts with a long tradition of falconry since 5th century AD from North Africa through the Moorish invasions which introduced the use of the hood. During this time period, the "Libro de los Animales que Cazan", a classical work on hunting, also known as the "Kitab al-yawarih" was written by an Arab falconer known as Muhammad ibn 'Abdallah ibn 'Umar al-Bazyar who lived during the 9th century. This was the first ever treaty to be written in Spanish on hunting and of which we have evidence.

Since then, falconry as a noble sport and art has continued to be present in Spain throughout most of its history and influenced its literature; indeed in Spain there are many classical works to be found which not only mention falconry but that have been entirely devoted to it. The most famous of these works is the "Libro de Caza de las Aves" by Pedro Lopez de Ayala (1385–1386). Other Spanish classical works which have also greatly influenced European falconry are the "Libro de la Caza" by Juan Manuel (which is the first book to be written on falconry in Spain, possibly inspired by "The Art of Falconry" of Friedrich II of Hohenstaufen) and the "Libro de Acetrería y Montería" of Juan Valles (1556).

However, in time with the introduction of the firearms, like many other European nations, falconry disappeared for a few centuries and was completely recovered by naturalist Dr. Félix Rodríguez de la Fuente in the 1950s. Felix had to consult antique literature and it was through the revival of medieval falconry that he based his own falconry. In 1964 de la Fuente wrote his outstanding "El Arte de Cetrería" which was to become a masterpiece and a book of great influence not only in the Spanish, but for serious falconers everywhere. For many, including myself, this book was to become the first step into entering the world of falconry and though written many years ago, I find that its techniques are still valid for modern falconry. Thanks to Félix, known as "the friend of animals", falconry started to flourish in Spain and Portugal after the 1980s and has become increasingly popular. According to official figures of the IAF, Spain is currently within the top five falconry nations in the world and enthusiasm is growing.

In Spain, as falconry was based in traditional or medieval falconry, only classical birds of prey were used and it was not until a few years ago that exotic species were introduced. These new additions to falconry were not at all welcome by the more puritan and elite falconers and as a result, the traditional peregrine falcons or goshawks were still the number one choice by most falconers or even apprentices. Then along came the Harris hawk, a bird which like the red-tail at first, was at the time rejected by most falconers. Now, funnily enough every falconer in Spain seems to have one or have handled one; without a doubt this is currently one of the most commonly used birds for falconry in Spain, particularly by beginners. Other "strangers" also arrived on the scene shortly after, like the red-tailed hawk, ferruginous hawk and Cooper's hawk although these have not yet quite taken on. Although some individuals had been imported into Spain both from wild-stock and breeding centres, red-tails in general were pretty much unknown before I published the Spanish version of this book and not too many falconers were willing to take chances with them. Since then, I am happy to say that the use of red-tails for falconry (and not just displays) has steadily increased but it has not yet reached the "boom" experienced by the Harris hawk.

As far as I have been able to see, out of the few red-tails available in Spain, nearly all of these are of small to medium size, and belong to the calurus subspecies. This is important to note, and this fact suggests that, until now, the training that has been provided to a great many of these birds has been improper. For example, this smaller subspecies will typically not be effective in hunting larger animals, such as hares, as some of the other larger red-tails.

The red-tailed hawk has its own distinct nature, and cannot be trained like a Harris hawk or a goshawk—and certainly not like a falcon. Although it is true that the best approach for red tails is a combination of techniques, as will be seen in the following chapters, one should take advantage of its strengths, rather than hinder their expression. With this taken into account, you can probably understand why Spanish falconers have been so reluctant to give this bird a chance, claiming that this bird just "doesn't work". The fact of the matter is that it has not been trained properly (or bred properly in many cases as there is a trend within some breeders to maximise red-tail clutches and therefore create imprints) and, consequently, its strengths have not been properly exploited. What we need to do is learn as much as possible about it. In this way, the red-tailed hawk can finally come to be properly utilized and we will be able to make use of all its potential.

There is a tendency to fly the red-tail exclusively from the fist, with the falconer attempting to induce the bird to capture its prey in the same manner as goshawks and Harris hawks. Yet it is often the case that the red-tail needs to attain higher altitudes in order to achieve greater speed so that it can capture faster prey. It can only do this if it begins its flight from trees, or from some other object high above ground. If this strategy is not utilized then, in a great many instances, the red-tailed hawk will not be able to capture its prey.

Portugal

Falconry is not widely practiced in Portugal although it is increasingly becoming more popular. The falconers who do hunt in that country tend to prefer Harris hawks to red-tailed hawks as low-flying birds. As far as I know, the only falconer in Portugal that has used red-tails actively in falconry (more than seven years) was Fernando Flores.

Falconry in Portugal – Fernando Manuel Flores

Falconry has been present in Portugal since the foundation of the nation in the XII century. However it has been practiced over all the Iberian Peninsula much before the foundation of the Portuguese kingdom, being introduced in Iberia through the north by Visigoths and by south by the Arabs. An ancient testimony of this appears in Asturias on the IX century.

The Law of "Almoçataria", in 1253, was one of the first regulations concerning hunting and falconry in Portugal.

The practice of falconry grew throughout the middle ages and it was during this time period that two Portuguese falconers wrote two famous treatises; Pero Menino, and in 1616, Diogo Fernandes Ferreira.

The golden age of Portuguese falconry was between the XIV and XVII century. During the reign of D. Fernando, a King known for his passion on hunting, many falconers were employed. He possessed more than 300 falcons and his hunting journeys on the regions of Santarem were known to have the company of higher hunting falcons. In his chronicles, falconry was described accurately.

These were the days when the Master Falconer was honoured and treated like a minister who depended upon his King.

In Salvaterra de Magos the Royal Falconry of Salvaterra de Magos was founded in 1752, employing Master Falconers coming from Denmark and Valkenswaard – Holland.

The Dutch kingdom offered a significant number of falcons and in 1780 the Royal Falconry of Salvaterra was employing 10 falconers from North of Europe with an equipage of about 30 Falcons. But there are records that show about 60 falcons imported in just one year from kingdoms such as Denmark.

This golden age of falconry declined later with the French invasions of Napoleon. The building was destroyed, and the new times posed a threat to Portuguese Falconry. Financial problems of the Portuguese kingdom and the increasing use of guns were, like in all Europe, the major causes for the decline of falconry.

The city of Salvaterra de Magos is currently making efforts to recover and rebuild the Royal Falconry.

Since the beginning of the XX century, falconry resumed once again in Portugal due to the efforts of 4 or 5 falconers.

It was only at the end of the '80s that new falconers began to appear in Portugal. They all united and in 1991 founded the Portuguese Falconry Association. During these years the association was made up by about a dozen members of whom only about half flew birds.

Nowadays, the Portuguese Falconry Association has 50 members, and about 20 are falconers that actually fly birds, mainly Harris Hawks and Goshawks but few also fly falcons.

3. The World of the Red-tailed Hawk

Fernando Flores and his female red-tail. Photo courtesy of Fernando Flores.

The law concerning falconry in Portugal is practically the same for all types of hunting, gun, archery, etc.

One is allowed to hunt between beginning of October and 31 of December for game like Rabbits, Hares, red legged partridges, and quails. The hunting of ducks is allowed until the end of February and doves can be hunted from the middle of August.

In order to hunt, a Hunting certificate is needed, plus an annual license and the annual insurance. Additionally, the falconer needs to pass an exam on falconry, which is mentioned on his hunting certificate. He also needs to show the correspondent Cites of the bird and an annual license taken each year for each hunting bird.

The hunting ground is divided into free land (now very scarce and without any kind of game) and private reserves divided in touristics, in which the game hunted is paid for daily or annually. Lastly we also have associative land, run by associations in which a fee is paid to hunt several days by year, or social hunting grounds, run by a local community, where an annual fee is also paid to hunt several days per year.

Hunting days are on Thursdays and Sundays, but falconers are allowed to hunt one day before, Wednesdays and Saturdays. On touristics one is allowed to hunt every day, depending on the annual planner of the ground.

The touristics are normally where a good density of game can be found, and it is difficult for falconers to compete with gun hunters. Touristics are normally run privately to obtain profits from hunting, and as usually each piece of game taken is paid for, gun hunters normally mean more profit than falconers, as the pieces shot daily are much more than those captured with a bird of prey.

In Portugal, the biggest problem that Falconry is faced with is the lack of game and things are even worse for falconers flying longwings. The most suitable land for flying longwings is found in the region of Alentejo, in the interior south of Portugal, with some open plains. This land is also the first choice for gun hunters and where game is commercially explored.

The remaining part of the country is mostly suitable for hawking with broadwings or shortwings, as it has many hills and dense forests.

With regards to captive breeding in Portugal, there about three domestic breeders and a large commercial breeder of birds of prey.

Hare Hawking with the Red-tailed Hawk

The Red-tailed Hawk (*Buteo jamaicensis*) is one of the most common birds of prey of North America and in its light form, also one of the most impressive and beautiful birds. It isn't yet very well known in the European continent, where sometimes and particularly in the Iberian peninsula it is compared with the Common Buzzard, (*Buteo buteo*). The truth is that there is nothing in common between the red-tailed hawk and the Common Buzzard, with regards to the quality and hunting ability.

Most Iberian falconers are not aware yet of the immense potential of the red-tailed hawk. Its use as a true falconry bird in the hunting of rabbits and hares should be considered a good choice for many falconers. Its use in North America is widespread and competes with the Ferruginous Hawk, Harris Hawks and even with Goshawks. Each species is conquering its own space, depending upon the favouritism of each falconer.

However in Europe it has not yet reached the same boom of the Harris Hawk, which conquered fans essentially due to its sympathy from many new falconers for its sociable and an almost dog-like character. The red-tailed hawk, is also easy to train but has more the temperament of an eagle, showing sometimes all its ferocity when someone gets between it and its food or quarry. Such character can become a little dangerous, especially when flying an imprinted bird. Even if at times the ferocity of an imprinted bird can have positive effects in hunting, I strongly advise, like with the true eagles, to use a parent reared bird.

When comparing the red-tailed hawk with other birds, such as the Goshawk, the most wanted bird for hawking, I would say that of course, the Red-tailed does not have the sprint and speed of the goshawk, but it possesses other qualities that make quite a difference: they do not show a sudden hysterical behaviour and it is easy to keep the tail feathers in good conditions. While it is true that they are not as sociable as the Harris, they make up for this in strength, particularly the females which have more powerful talons. This strength can make all the difference when chasing hares. Red-tailed hawks also have the speed needed to catch the fastest and most powerful of hares.

The choice between these three species of raptors, in my opinion depends on several factors, such as the availability of quarry, daytime and available hunting land. For rabbits, I will choose a Harris or male red-tailed hawk while for hares I prefer the female red-tailed hawk. For experienced falconers with a lot of patience and willingness for sacrificing a few tail-feathers should choose a Goshawk.

For about 6 years the red-tail was my first choice for hunt hares, it was a truly enjoyable experience with my female at 1,200 grams (42.3 oz) of hunting weight. In those years her margin of success was so great that I never thought of asking her for more.

During her first years, I put her onto rabbits, and the ease of introducing her to hunting them was amazing, almost instantaneous. After one or two seasons on rabbits, I found a good hunting ground with hares and a lot of opportunities, and again it was easy to introduce her to them. She became a deadly enemy for the hares. At times, I tried again to hunt rabbits. I would have to do it after she had caught one hare and not before. I believe that if we ask a bird to catch rabbits, and afterwards we ask it to hunt a more challenging quarry, it can establish some preferences for the easy quarry. At least by hunting in this way, I have no concerns.

After the moult, when the bird is normally too fat, and lacks fitness is important to get her gradually down to her hunting weight, flying as much as possible in order to build up muscle and loose all fat, before getting her to chase wild quarry. It is very frustrating for a bird that goes hunting directly from the moult, to not be prepared to deal with wild and muscled quarry which is used to evade predators all the time. Taking unprepared bird hunting is like teaching her that the quarry is not for her and will only result in the bird getting demoralized needlessly after some failed flights.

Building up muscle in a red-tail takes a little bit more time than muscling others birds of prey such as falcons or goshawks. Somehow it is like muscling an eagle. It is a large and heavy bird and to increase her fitness I used to fly her in hilly country calling her successively to the fist up hill. Flying up hill demands a lot of strength from a bird. The distance was increased when she arrived with less effort, and so was the inclination of the ground and difficulty. If at the beginning she got tired too quickly and couldn't reach to the fist, I decreased the distance or waited until she recovered her breath and started flying climbing again the hill towards me. That's the quickest way that I've found to muscle her before hunting.

The second part of acquiring fitness was in true hunting that gives each day more flying speed. At the beginning, during the first flights, the success wasn't anything to be amazed at, and when she did get a quarry I would just let her eat her quarry at will. Just a quarry at time is more than reasonable in its early hunting days. In time, the catches will become more frequent, almost daily, and if at the beginning it was necessary in failed flights to get at least a quarry taken, the situation was starting to invert. At the mid-season the bird was acquiring a very big rate of success, and towards the end there were more flights resulted in quarry taken than in failure. Back then, the number of hares per hunting day was increased, and it was not difficult to reach the middle or end of the season catching about three hares in a morning or evening of hunting.

I just try to be cautious in days where there is a lack of quarry. If the bird gets its first hare at the end of the evening and the probability of finding a second one are not too high, I will let her eat at will on her first kill.

The hunting strategy was decided on each day. On windy days I try not to cast the bird against the wind. This is particularly important on broad wings. Hares on these conditions have a strong advantage, and the bird will only have success in close slips. The hare knows it and will run against the wind. On the other side, flights with strong tail wind are also not recommendable, as the bird will achieve an enormous velocity, but loose control and manoeuvrability, not being able to turn easily. In these situations, normally the hares pursued will stop suddenly and change direction, evading its predator with great success, or will simply jump on the vertical, putting the bird beneath, grabbing the ground. The bird at full velocity will not be able to stop with the same efficiency while continuously being pushed by the strong wind at its back. Under these conditions the cast must be searched by wind blowing by the side. Another thing that happens frequently when we walk with our backs to the wind is that this causes the hares to run very far away, since they hear easily hear us approach. We can get closer to the hares on windy days if we walk against the wind, taking precautions for the flight being made by the side of it.

On hilly country, downhill flights are preferred, so I walk with the bird on the fist along the upper edge. This allows easier slips, with longer chases, and we dominate a lot of ground. Casts uphill are a little bit more difficult and the success depends on the fitness of the bird, and the distance and advantage taken by the hare should be less, depending also from the inclination and the wind force. All these circumstances should be measured before the flight.

Usually hares prefer to lie down in a sunny ground descent, protected from the wind; these parts of the country in winter mornings offer a warm place when the sun arises after a humid and cold night. A Field of freshly-cut corn or low vegetation is preferred by hares, like for example fields with low thorny vegetation that offer some shelter. They like to keep some distance from places with a

3. The World of the Red-tailed Hawk

lot of presence of cattle such as sheep or cows. When they feel pressured with hunting of guns, they will tend to wait a longer laying down until the last moment, when they feel threatened or are about to be discovered. If there are no more gun hunters on the hunting ground, they will tend to be running around more and be more confident, frequently interacting with each other.

After some time we are able to train our vision to locate hares even when they are laying down in their "beds". This is when we should position the bird in a convenient place and direction and throw a little gravel to the place where she is, in order to make her run. Since we give hunting its own dignity and there is nothing worse than to cast a bird to a hare that stands quietly in the ground, all efforts must be made for the flight to be fair both for the prey and predator, and at the same time result in a beautiful scene for the observer, the falconer.

Catching the quarry or losing it should really be one of our last concerns. At times, an experienced bird learns to spot its prey when it is laying down, normally after the hare makes a minor movement while being still. The bird usually sees it before we can. The bird of course will always choose the easiest way. Usually, it will fly off the fist directly to the ground, causing in my opinion a bad flight. It is in our judgment to decide to let the bird go only when we also spot the hare running. On the other hand, sometimes hares flush and run just behind us, the bird sees it but we hold it secure, not letting her go. As soon as we realise this, the hare has just gone, the bird is hanging and bating and we have lost a good chance of having a good flight. Everything can happen. In only a matter of seconds, we must decide to let her go or to keep her on the fist. Sometimes we make the right decision, sometimes we don't.

When flying from the fist, flights at game are difficult, more challenging, more fun, more sporty and perhaps more beautiful. Seeing amazing pursuits and being able to control everything, the strategy as a whole, we are present every time and interact more when hunting. When the bird fails, getting an amount of grass in between her clutches instead of a hare, we can just call it or recover it directly from its position. The bird is generally more obedient as it understands that we are necessary not only to offer it a perched and moving place, but because sooner or later we will flush to it again another opportunity.

Going back to the rabbits, it is important from the beginning to understand that they are an easier quarry, and also more attractive for the bird. If one day I decide to go rabbit hawking, if possible I would choose one day just for the rabbits. If you are trying to make a new bird and introduce her to hares, then forget about rabbits. However, if the bird learns to catch hares and is experienced, choose the rabbit at the end of the day, after it has already had a kill on a hare. Just be sure to not let her fly at rabbits in order to "save the day" and make up for a bad day of several missed flights at hares. That is just like teaching the Bird: "OK, if you are not able to get some hares,

The Red-tailed Hawk

Female red-tail working out flying to the fist. Photo courtesy of Fernando Flores.

let's try to get something easier." Next time round, she will certainly prefer the easier quarry, rejecting the most difficult. We are just simply conditioning preferences. Do not forget that securing a hare demands a lot from the bird. Adult Hares can have weigh in average between 3 and 4 Kg (7–9 lb) and when they feel grabbed, they put up a strong fight, kicking all they can and this can cause some jumps in the air. We must run and help the bird as soon as possible,

3. The World of the Red-tailed Hawk

*Fernando Flores, his female red-tail and a freshly-caught rabbit.
Photo courtesy of Fernando Flores.*

securing the legs of the Hare and ending with its suffering as quick as possible, normally pushing suddenly the rear legs when the left hand secures the back of its head, breaking its neck immediately. It is a quick death and will end the suffering of the hare. The bird normally learns to secure its prey from the head, and this is the correct way. The bird quickly will learn that when we arrive and make in, we are not a competitor for the food but just help in securing its prey. All we have to do is let her eat from the prey all that she wants, normally one or two front legs. Don't worry though; you will still get plenty of hares.

Falconry Events

Competition with birds of prey is new and is growing fast between the Falconry community. By some how it's a result of the increasing numbers of falconers in the last decade together with the increasing lack of ground and game availability. Being not the true meaning what falconry is, chasing wild quarry with birds of prey, competitions arise because there are a lot of would be falconers that flying daily their birds of prey don't have available hunting places, being the alternative to train and maintain the bird flying and chasing, flush domestic and controlled quarry such has pigeons and partridges. Being a place where normally in tree or for days a significant number of Falconers are together, gives the opportunity to change conversation and establish new relations between each others. Helping to increase the number of participants and visitors, a market was established, with the presence of manufacturers of falconry equipment, books and videos, breeders, artists and every related falconry merchandise. To motivate the participating the prizes are made a temptation. All these people attracts also to the regions where the event is being made regional sponsors, and even support from the public and local administration.

Competition is sometimes seeing by purists falconers with some negative and exposing effect, but if organized carefully it should be seeing not like a true falconry journey but like a "fiesta" and an annual meeting point, for the fans of this art where everyone can pass an excellent weekend. Of course that all the flights must be seen as a controlled environment with controlled quarry always present at the right time and flush, all that in true hawking never happens.

On the last years, in Spain the numbers of competitions is growing exponential, being made all over Spain in different provinces.

I personally have participated with a Red-tailed to hares and with a Peregrine Falcon to pigeons on the Sky Trial, and simply I think more enjoyable be present without birds than travelling about 5 days, getting in a Hotel with the birds, difficult in getting some fresh food. For those that have to travel from far away I just think it's not worth it. Birds are better hunting with all they need, and be present just during the weekend without a bird is less stressful and more enjoyable. To those who live near the event, things are much easier.

Croatia

Falconry in this country has experienced resurgence after this nation's War of Independence, which ended in 1995. Still, the sport is relatively new there. During more than forty years of Communist rule, the sport was prohibited, probably because it was considered "aristocratic".

It is estimated that there are a total of some 20 falconers in Croatia, and the government does not require a hunting license—in contrast to Italy and other European countries. However, greater efforts should be made to rehabilitate birds of prey, and to treat them when they are injured. Knowledge about birds of prey should also be disseminated to the wider public. Marco Calistri recently visited a centre for birds of prey run by his friend Emilio Mendjusic. This centre—*The Sokolarski Center*—is part of the IAF and is located in Dubrava, some 6 km from Sebenico. This falconry centre is located in the middle of a beautiful pine grove.

3. The World of the Red-tailed Hawk

The Red-tailed Hawk

*Buteo jamaicensis harlani, juvenile captive-bred individual.
Photo courtesy of Dennis Lorenz.*

4. Equipment

Before we even think of bringing a red-tail "home", and after seeing if it is viable and if we will have the time to properly care for her, we must sort out the future lodging or facilities where we will keep our bird. Will it be necessary to build a special facility for housing our hawk outside or can we already make use of an existing facility and perhaps adapt it? These and many more questions should already find a reply in your thoughts.

A good idea for housing would be an outside mews or "weathering" similar to those used in the UK and United States, as red-tails accumulate a lot of fat reserves and tolerate extreme temperatures which translates into making it a little easier to keep its weight as it should be. We should however, make sure this housing is safe from any intruder (including wildlife such as foxes or cats), away from direct draughts and with a covered area to avoid rain.

Such type of lodging should also be designed keeping in mind the security of the area we live in (possibility of burglars or wildlife attacks?) and many times, it will not be sufficient to construct it solely using wood, but perhaps the best solution, as was my case, could be of constructing at least in part of cement and or/bricks. In any case, it should also be built in a place where our bird can have a nice view to stop it from getting bored, as if we put this directly in front of a wall, it could well start to pluck its feathers out of boredom. It must also be easy to clean, solid and resistant against all weather conditions and act as a shelter from both rain/snow and cold. That's why it should never be placed in a very windy area. It must also make use of a strong wire that will avoid foxes and rats from digging in and getting inside our hawk house as in such a small aviary our bird will be tethered and therefore defenceless against such predators.

It is not necessary for it to be huge but if it is made of wood, it should be as solid as possible to withstand all weather conditions and possible break-ins by wildlife. Ideally, it should have three sides or walls protected by a forward-slanted roof that will allow the rain to pour off it. The last wall should be made of strong wire and act as window that will hopefully offer some nice views to our red-tail. The ground can be covered by gravel or something that is easy to clean with a garden hose daily.

Needless to say, in these types of small-sized lodgings it is recommended our bird be tethered and not free-lofted as being so small, it could hurt itself if it was flying free. That is why safety and durability are essential here to protect it from unwanted situations. The construction should in fact be as simple as possible though.

Another alternative for those who aren't so handy with DIY is to buy a module as the one we can see in the photograph below, made in tough steel which we can place (with a cement base to avoid digging of predators) in our chosen place in the garden. We can then cover the ground with gravel or small pebbles to make it more comfortable and easier to clean. These modules are usually sold in big DIY stores and can sometimes also be purchased online.

The Red-tailed Hawk

This practical steel gardening-module can be easily assembled and with a few small modifications, become the ideal hawk house for a red-tail and other birds of prey, such as owls. Photo by the author.

External weathering where birds of prey can be kept during the day for sunbathing and taking a bath. It is recommended that this type of weathering be built in an area where we can have our birds of prey worry-free relaxing during the day, especially if we are not going to be there to keep an eye on them in order to avoid possible dangers (other birds of prey, cats, wildlife, etc.). Here we will place our blocks and bow perches, and if we place several birds of prey (always tethered), we will make sure they have enough distance between them to not each and start a fight. We must also keep an eye out for extreme temperatures, both in winter and summer as our birds may suffer from hail or rain if they do not have a sheltered area or in summer, from dehydration and excessive sun exposure (death) if we also don't provide a shade area as the one we can see above. Photo by author.

4. Equipment

Purpose-built chambers or mews for raptors, with brick and cement. Very safe and ideal for greater intimacy during the moult and breeding season. Photo by the author.

If we already have a place inside a building, aviary or barn for example where we can house our hawk, with a few adaptations, we can make use of it. It is important that it has a big enough space around it and that this is relatively calm so that it does not stress out and damage its feathers. However, this is usually impossible to avoid at times, especially when we have a new bird and we have not trained her yet as they can be quite uneasy then.

We can also purposely build a cement or brick construction similar to a breeding chamber (which could be used precisely as that later on) with a partly open roof (covered with mesh wire). This chamber will serve several purposes if we build it well and add natural elements such as sticks and one or two shelves. It will not only act as a hawk house (always tethering our bird if it is not trained, once trained we can move on to "free-lofting" it) but also for the moult and breeding seasons. If we are going to build such a construction, we should especially make sure that it is in a calm place, away from humans, loud noises and interference as this could cause stress both during the moult and breeding season with catastrophic consequences.

Essential equipment for daily handling

Before we decide to have a bird of prey and after reading up on it and hopefully attending some sort of course to prepare for the new-comer, we will also have to buy some basic material for handling our birds of prey daily and during training. It is possible to make some of it at home if we are handy, especially if we have gone to a course, such as jesses, hood and other leather wear, but we will have to buy many other things. The following is a recommended list for beginners of what you should have:

- Hood (we can make this ourselves and if not, also buy it, but it will always fit better if we can have it made to measure. I have included a sample hood pattern in this chapter for this purpose)
- 1 double-thickness falconry glove (please mention that it is for a red-tail when you buy it as it needs to be quite tough)
- Jesses (training and field) and leash
- Swivel
- 1 bow perch, block perch or screen perch (the latter not recommended for new birds)
- Scales
- Falconry bells
- Squirrel Chaps
- Lure (both dragged (or rabbit lure) and swung (winged lure)
- 1 hawking bag or waistcoat
- Telemetry system: Transmitter and receiving unit in proper working order with batteries!
- "Giant hood" or transport box
- Bath pan

It will also be useful to have certain tools like whistles (for training), veterinary nail clippers for snipping off the point of the beak and talons if needed, grease for jesses, tools for making hoods, imping needles, etc. Please refer to the vocabulary at the end of the book for a more complete explanation of some of this equipment.

If we decide to be lazy and not attempt to make anything ourselves but rather purchase by mail order or internet most of our falconry equipment, we must mention the size (weight) and sex of the bird, species and if possible subspecies (as physical characteristics will vary) of our bird. There are quite a few subspecies of red-tail as we have seen in the previous chapters and they don't all have the same size or shape, plus females and males will also differ. The sizes usually used for most red-tailed hawks are large (males) and extra-large (females) although this may vary depending on the individual. This is also one of the reasons why it is good to know how to make a few basic essential pieces of equipment, just in case. Please check the rest of this section plus Appendix of this book for additional material tips and patterns and the DIY section of my falconry website at **www.yarakweb.com**.

Hoods

Hoods are not only used with falcons but can also be used with hawks though perhaps not in the same way. In Europe, hoods are not commonly used with hawks and are usually only used with falcons. However in the Unites States, red-tails flown are usually wild-caught hawks who have not been bred in captivity and it can sometimes be easier to get them to use the hood. Nevertheless, though red-tails don't usually need hoods, it is a good idea to introduce them to such equipment as they can be used for transport in stressful situations, or when we cope their beak, talons, etc. Surprisingly, many of them will

4. Equipment

be good hooders, as was my male red-tail who can see in the following picture, about to get his beak coped. In this way, hoods can be extremely valuable. Funnily enough though, some red-tails with time (like my own) will turn out to be enormously patient and actually allow you to do such things to them without a hood, just being perched on their usual block or perch, but this will not be the usual case.

I do not however, believe in using hoods with these hawks on a daily basis as you would with falcons (if they are bred in captivity) as they will always react better and actually be a lot calmer without them.

"Bandit" with his new home-made fitted hood. Photo by the author.

The Red-tailed Hawk

There are many different types of hoods available and we can even develop our own designs. Hoods have always traditionally been made sewn (still are in Europe) but in the United States and Canada, they are usually glued. The size of the hood will not only depend on the subspecies and sex of our bird, but will also vary within individuals of the same sex of a given subspecies. Please have a quick look through the subspecies section again if in doubt, as for example the head of a female B. j. calurus may be the same size as a male B.j. borealis or kriderii. Generally though, females will use larger hood sizes. When buying a hood, we must always choose comfort first instead of choosing a hood that may appeal to us simply because of its aesthetics. Once we do choose a design and get our bird used to it, it will be quite difficult to introduce it to a different style so do choose wisely.

Next, I have decided to include a hood pattern that was designed and made to measure for my red-tailed hawk and which appeared in a previous

Hand-crafted hood made by English falconer and equipment maker Ben Long. This hood is actually quite large, as we can see in the photo, and made for eagles, but due to the shape of a red-tail's skull which is very similar to that of an eagle, we should opt for these types of hoods with a few minor adaptations for our birds with regards to size. Photo by Ben Long.

4. Equipment

photograph in this section. This hood pattern is for a large eastern male red-tailed hawk, which is usually much larger than a western (calurus) male or female red-tailed hawk and the skull can also be a little different here. You will possibly have to adapt and modify this hood pattern to the size of your own red-tail. In the Appendix, at the end of the book you will find other patterns for falconry material as well as online, in the DIY section of **www.yarakweb.com.**

Anglo-Indian Adaptation - Hood Pattern For Red-Tailed Hawk

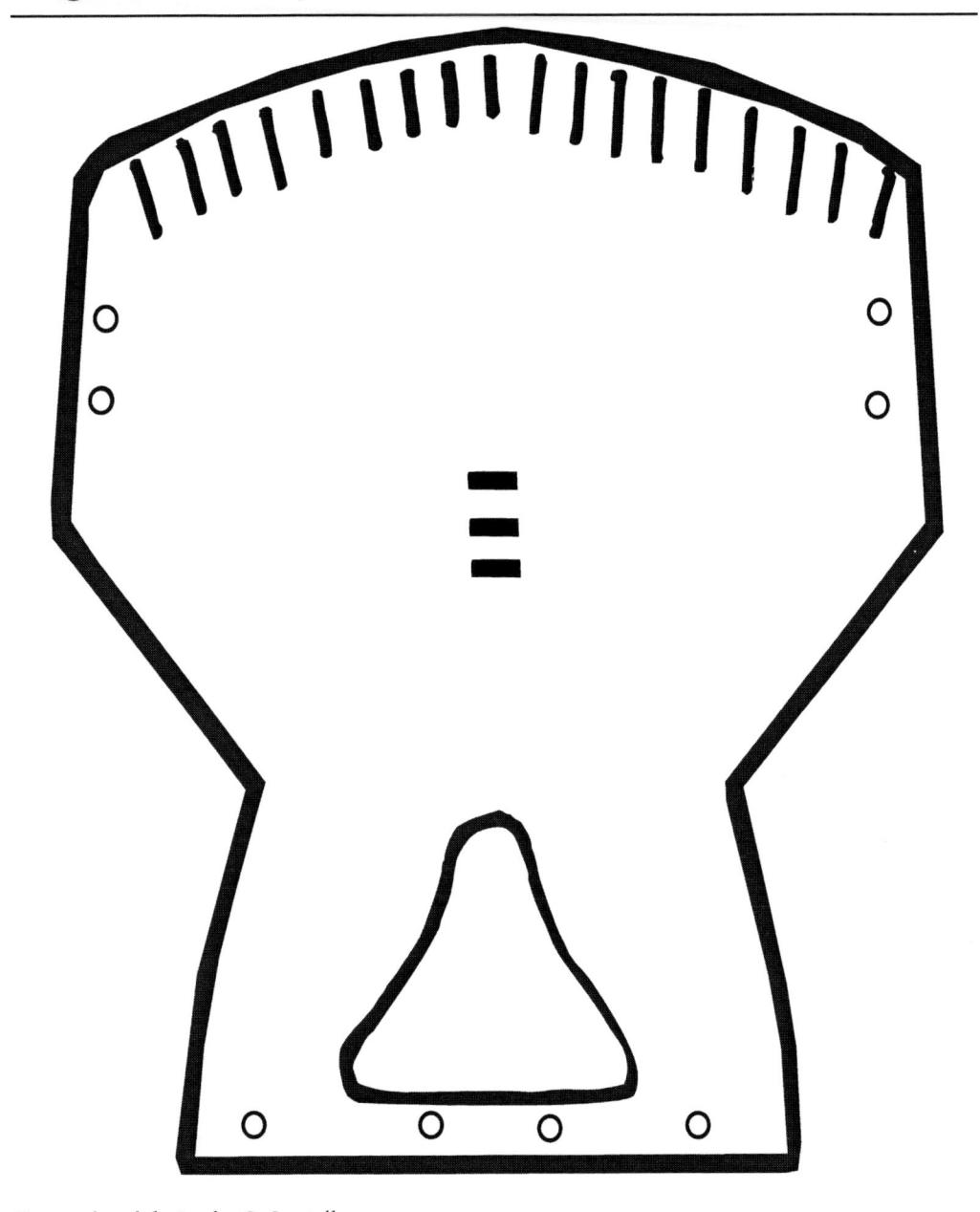

From a hood design by G. Santalla

The Red-tailed Hawk

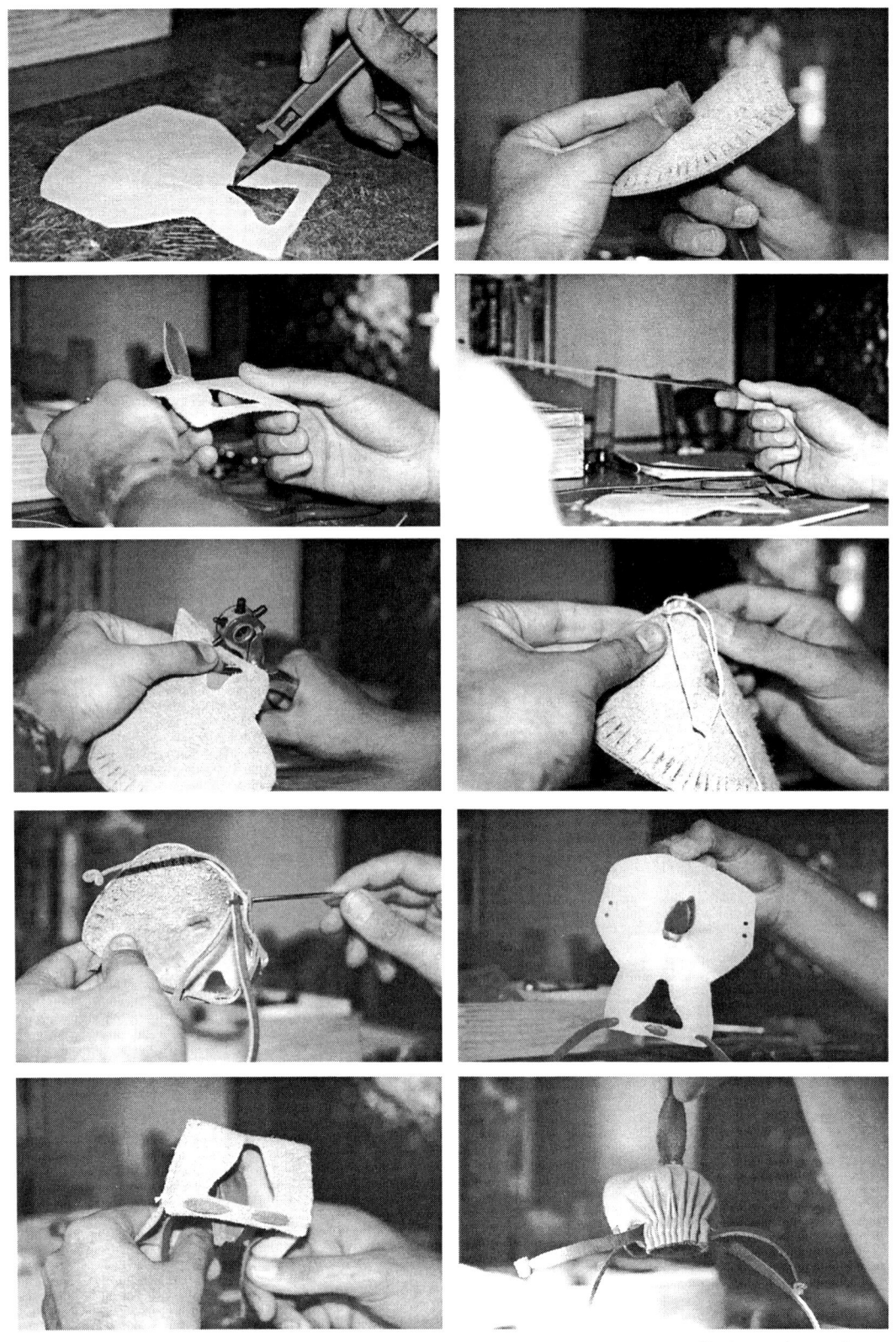

All photos by author.

Anglo-Indian Adaptation - Hood Pattern For Red-Tailed Hawk

This is an adaptation of an Anglo-Indian hood which uses the closing feature of pleats, causing the leather to gather at the back, in the typical way of Arab hoods. It is lightweight, easy to make and provides wide comfort in the eye area.

You will need:

- Soft semi-rigid leather (any colour)
- Scissors
- Cutter
- Sewing needle (for leather)
- Strong nylon string (must be waxed first)

1. First make a copy of the hood pattern, cut it out and place on top of the desired leather. Draw around it and then cut with scissors, save the beak area (this must be cut out with a cutter - make sure to use a cork board underneath! The decoration element or top of the hood must be cut out the same way, also using a cutter.

2. Cut two long strips of about half a cm (0.2 in) wide for the braces.

3. Insert the top bit through the three middle cuttings (see photos below).

4. Sew (in the US, hoods are usually glued) the hood with the inside out, getting as near as possible to the edges but without tearing the leather apart, ending with a knot. Sew both sides.

5. Flip the hood so the outside is now correctly on the outside (usually for comfort, the nicer bit of the leather is used inside) and insert the braces through the openings at the back. Wet the inside of the hood and shape it, also widening it a little with your fingers, especially around the eye area. Leave to dry naturally.

Falconry glove

When handling red-tails, it will be necessary to use a good quality double-thickness glove to protect us from their powerful feet, which could seriously hurt us if we didn't use such a glove at times. Falconry gloves are also useful as they serve as a cushion where birds of prey may rest for long periods of time without hurting themselves, like for example when manning as red-tails will require longer than perhaps other birds of prey.

I do not recommend picking up any bird of prey without a glove, let alone a red-tail! Not just because of its powerful grip or sharp talons but also because we may actually hurt them.

Going back to the types of gloves, there are many colours, styles and leathers used for these. One note of advice here, as with hoods, birds of prey will get used to a particular colour or style, so try to always use the same type of glove in order to avoid problems.

My personal preference if for deer or elk-skin double thickness gloves (with a nice warm comfy lining inside) which are more comfortable and less rigid to handle.

The Red-tailed Hawk

These are also quite handy in winter when our fingers are also quite cold! In Spain and many parts of southern Europe this exact type is quite hard to find and though there is a lot of choice with regards to gloves and some may even be beautiful, they may not be adequate for falconry or for the bird we intend to fly. I have always chosen to buy these in the UK and my favourites for years, like most other falconry equipment, have been those made by Ben Long.

With regards to the glove, the design in itself is not so important, but double-thickness is necessary for red-tails. Additionally, our glove should also come with a security clip in order to "avoid accidents" and if it doesn't we can just make this ourselves. We should always tie our red-tail securely to our glove when carrying it on the fist and this clip can come in handy as if we do trip, it will it will stop our red-tail from flying off our hand. Imagine this situation in a newly untrained bird: it could be disastrous. Last but not least, the tassel is also essential not just for hanging our glove, but it can help to clean ourselves up a little when out in the field. This was one of its functions in the old days although now it's usually just seen as a decorative element and just used to hang our glove.

A glove for red-tails should be of a medium size and length as if we get one that's too long (like those used for eagles), it will be uncomfortable and make handling difficult. A shorter one will be insufficient and cause injury. The glove should feel quite snug, not too loose or too tight, just right.

Even if it is made out of leather, it should also be cleaned out regularly but we will NEVER use soap and water for this as this would ruin and harden the leather. Once we have cleaned our glove superficially with a moist kitchen wipe, we can use a brush and use a little leather grease. If for any reason, the glove were to get wet, we should let it dry naturally (just like hoods when they are moulded into a shape) if possible near a heat source but never use anything to actually dry it.

In any case, if you fly your bird as often as you should, you will most likely need a new glove quite frequently and this is sometimes the better option. I used to handle many birds of prey every day and would use my glove till the last moment, which used to be a maximum of six months for me.

Anklets, jesses and leash

It would be a good idea to also learn some minimal DIY falconry equipment-making skills so we can at least know how to make some anklets, a pair of jesses and a leash. Although these items can usually be bought in most falconry stores, even online, we could have an emergency situation suddenly and need to have another pair at hand immediately. If we can quickly make in a few minutes another pair, it would definitely be handy.

Jesses are simply two leather strips which go through the leather anklets so that the falconer can hold the bird or attach the leash. Traditionally, these leather strips were made out of dog

4. Equipment

Ben Long is an excellent falconer and equipment maker known for decades not only in the UK but worldwide for his quality falconry equipment, particularly his hoods, (see photograph).
His gloves are amongst the best that I have used and are attractive, comfortable and practical; once you try them, like me, I am sure you will not want to use anything else!
Photo courtesy of Ben Long.

Beautiful hand-painted falconry glove of "Bandit", the author's male red-tail by Italian miniaturist and wildlife painter Claudia Panniello, a good friend and great person. Claudia works with different materials, always under commission, and her work is really unique. Photo courtesy of Claudia Panniello.

The Red-tailed Hawk

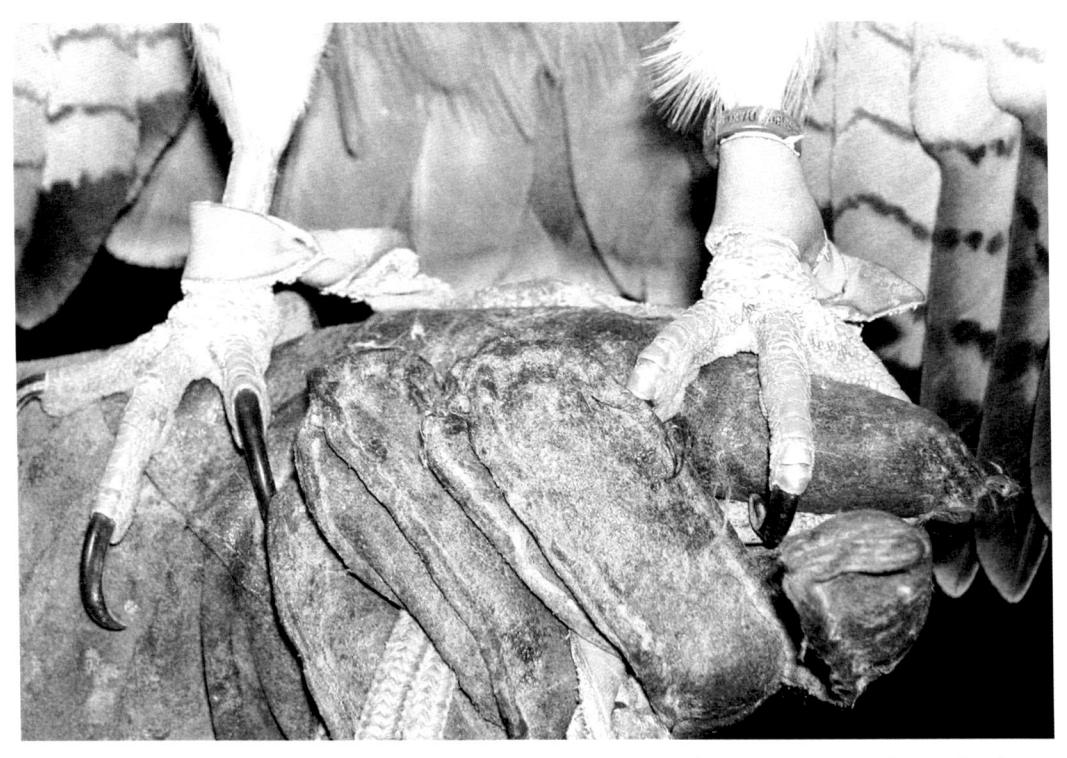

Above, we can see the author's red-tail "Bandit", with traditional jesses that the author made herself, well-greased to avoid her biting them (they really dislike the taste!) and these proved to be quite strong. These can be used always, although they are especially recommended in the initial stages of manning and training. Red-tails, just like Harris hawks and other hawks, are better suited to anklets and Aylmeri type jesses (with interchangeable jesses for the field and when it is tethered. Photo courtesy of Francisco Solano.

Left, we can see a red-tail on a bow perch made by Ben Long. This design is ideal for red-tails, as it allows them to rest adequately avoiding damage to the feathers and if we do not wish to purchase it, it is quite easy to make at home. Photo courtesy of Ben Long.

leather. Modern jesses are of many types of material including parachute cord and various braids, although the best material, is kangaroo leather.

I have include a standard pattern for jesses, both traditional and Aylmeri, and anklets for a red-tailed hawk at the back of the book (see Appendix). Note that the length of the jesses should be a little longer in the beginning, when we are beginning training as it will be easier for us, but that in any case, these must always extend considerably past the tail.

Jesses can be of any colour we like and preferably of the best leather we can find (kangaroo leather). They must be wide, strong and properly greased for better elasticity and to avoid being bitten by our hawk. Leather grease has an awful taste and our hawk has ever thought of biting through her jesses to get lose, once it gets a taste of them, this idea will be radically erased from its mind forever. This is exactly what happened with my red-tail "Bandit" who during the first few days tried to bite through them. He has never touched them since.

When we first make our traditional jesses (or buy them), we can use a longer, wider measure, just like I did (see previous photo), as this will make handling during training much safer and easier. Once training is over and before our red-tail flies free, we can change these for an anklet and Aylmeri system with interchangeable jesses. This will allow us to use slit-free field jesses when hunting which can help avoid many accidents as these will not get caught in tree branches or bushes. I also recommend starting off with the more traditional jesses, as if we have no experience in making falconry material, it is easy to make a mistake and these may not be properly closed and may open up. There are many "horror" stories to be heard about falconer's whose birds "broke free" when using this system after only 3 or 4 days. This is something we really want to avoid, especially with captive bred birds such as the ones used for falconry in Europe. These birds will not be trained and thus, will not know how to hunt properly and could starve to death if they break free and get "lost". In falconry, we must always think ahead and careful planning can save us from a great deal of headaches and also save your bird's life.

The leash was also traditionally made of leather and we can still see many falconers today using these types of leashes. Modern leashes have taken many forms and many materials are used including those used for hiking such as nylon (see photograph). A leash is what attaches the bird to the perch or falconer's glove. Again, quality is important here and it must resist constant bating and pulling by our red-tail. We can also easily make a leash at home in a few minutes. All we need is to buy a little hiking cord, have a small piece of leather at home for making a button (knurl), a lighter. We make a knot at the end of the leash, thread the leash through a small hole made in the middle of the piece of leather and melt both ends of the knot is melted with the lighter, to keep it from fraying, and a little super-glue.

A selection of swivels, from left to right: different variations of a figure-8 swivel and a fishing swivel. Photo courtesy of Francisco Solano.

Swivels

Swivels have replaced the old-time varvels and must be rotating. They are used to avoid entanglement while a bird of prey is tethered and connect the jesses with the leash. There are many types of swivels out there but the most adequate for a red-tail would be a figure-8 swivel or sampo swivel. These can be made of many materials but the best is titanium.

Block perch, bow perch or screen perch?

In its natural habitat, the red-tailed hawk usually likes to perch on tree branches and telephone posts near highways. For this reason, the best "perch" for a red-tailed hawk would be one that is most similar to its natural perches and would therefore avoid damage to tarsi and feet.

We have quite a few options for red-tailed hawks. We can either use a solid block perch (this must be quite heavy though, as many red-tails can drag traditional large block perches), a bow or ring perch (these being the most recommended) and also a screen perch although generally, I would not recommend this for red-tails as due to its nature, it could damage its feathers through constant bating and if it is near other birds, it would be trying to reach out for them and this would be unavoidable.

Scales and weight management

A good weighing scale is another investment we must also be prepared to make and will be a key piece of equipment for proper training of our hawk. Without a correct reading of our bird's weight, it will

4. Equipment

be impossible to carry out an efficient and proper weight management and this is basically the key to successful falconry.

There are many different models out there (both traditional and digital) and what we choose is up to us, but it must be accurate. We will probably have to adapt it for our hawk and make a small T-perch on the base so our red-tail can be perched naturally when we weigh her daily.

Our bird will have to be weighed daily, always before eating/flying, at the same time of day and avoiding any movement while doing so in order to get a correct reading of the weight of our hawk. We must also check the scale regularly to see that the measurements are accurate.

If we have accustomed our bird to be weighed with the hood on (though if it has been bred in captivity, this should not be necessary) we should continue to do so. Once manned and trained, even during the process of training itself, will relate getting weighed to getting food and will be happily anxious to jump on to the scale. Our hawk will have learnt that after that little jump, she will get food in someway and will not be problematic or bate, unless we have food near us and she can see it. Red-tails are very intelligent birds, much more than we think, and will always know where the food is. If they see anything they want near them, they will not hesitate to bate strongly towards it, as if their life depended on it. This could be another bird, food, or even a person. That is why when we weigh them, we must weigh them tethered to our glove just in case. I also don't recommend anyone to put food in their pockets, for the obvious same reason – ouch!

After we weigh our hawk daily, the weight should be noted down in a log book or register (I have included a sample sheet in the Appendix section which you can photocopy or even create and adapt plus print for this purpose), as well as the food eaten or game caught, temperature and weather conditions and what exercise or training we have done that day plus anything else which you may find useful. This daily log book will be of great use in learning to establish a diet and seeing which food works best for your red-tail and will help you maintain a proper weight management for success in the field.

Falconry bells

Before the arrival of telemetry, bells were always used in falconry and played an important role alerting the falconer to the bird's location in the field. These bells were small and usually made out of brass, silver, nickel or stainless steel. Traditionally, two-toned bells, each with a different sound, would have been used for a bird. Even with telemetry, the use of bells is always recommended.

These are still quite necessary in falconry for most birds of prey (save owls) and can be an additional guarantee for recovering your bird if it gets lost as a transmitter could break or fall when hunting. They can be attached by using a bewit to the tarsus of our hawk or to the central deck feathers, just like a transmitter. The latter is quite a good idea as if we also use chaps when hunting, our red-tail could find that there are too many things in the way.

The Red-tailed Hawk

Falconry bells traditionally hand-crafted in Spain by José Manuel Gamito, Estepa, Seville. Photo courtesy of J.M. Gamito.

Falconry bells are also traditionally hand-crafted in Spain and the most well-known type used here are the "Filigrana" falconry bells made by artisan **José Manuel Gamito from Estepa, Seville**.

Squirrel Chaps:

This is quite an innovation in falconry and rarely seen in Europe although falconers here would also learn to appreciate them if they used them. Similar to a type of leather anklets, they cover part of the tarsus and feet of a hawk and help protect her from bites or wounds. These are used in the USA for squirrel hawking and can also be a great aid when hunting hares or rabbits, as a bite by any of these can cripple a hawk for life.

There are two types of chaps that can be used with red-tails: fixed or removable chaps. Chaps will also have falconry bells integrated into their design.

It is quite easy to make a pair of simple chaps and we will need the following material:

- Strong and flexible leather (for example the same type of leather that we used to make the falconry hood in the previous pages)
- A pair of falconry bells
- Decorative metal studs. Any shape or colour can be used and are quite useful because squirrels seem to like them. If the decide to take a bite at our hawk, they will be first biting the metal stud and leather chap so will not likely hurt our bird
- Grommets (like the ones used for Aylmeri anklets)
- Leather glue
- Some thread (the same type of thread that we use for sewing hoods) and sewing needle
- A pair of batteries (1.5 V size) to be used as reference for diameter and size of tarsi, although if our red-tail is quite still, we can be making these made to measure. The batteries will also be used to mold and shape the chaps

Instructions:

1. Draw the chap patterns at the end of the book onto leather and cut out.
2. Place the diamond/stars studs, etc on pattern A with glue and then glue to pattern B. Leave to dry while rolled around the 1.5 V batteries.

4. Equipment

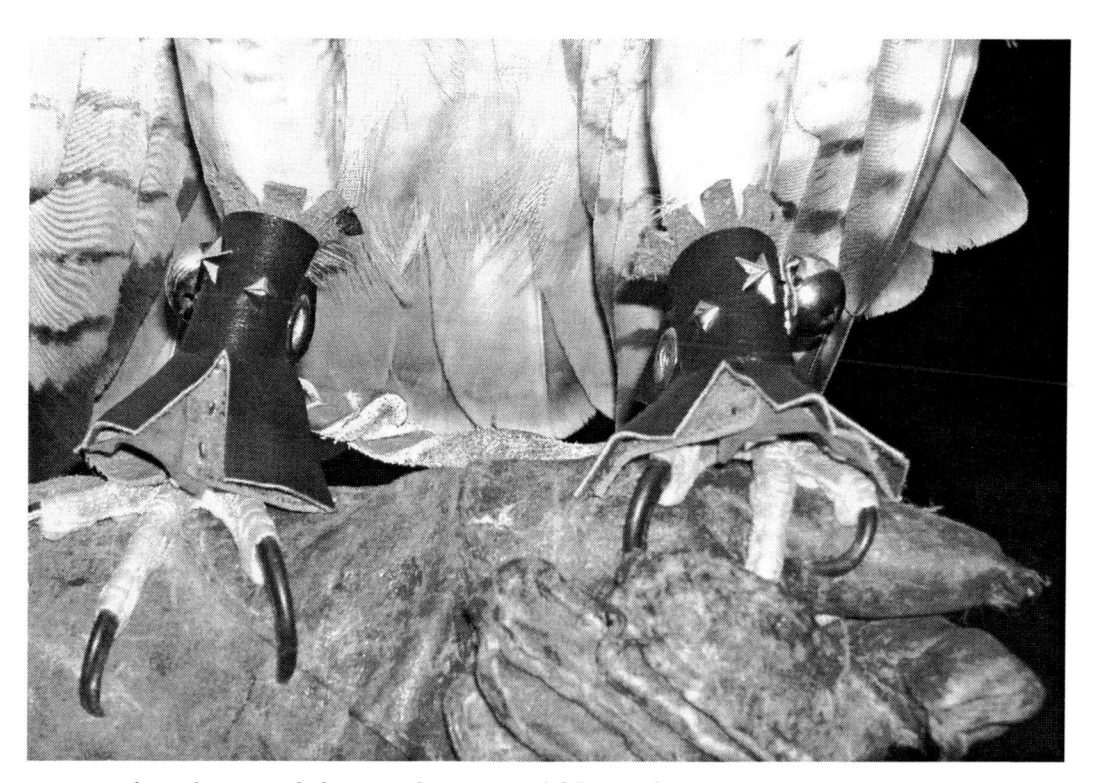

Hand-made squirrel chaps with integrated falconry bells inspired on the Gary Brewer models, similar to those used for squirrel hawking in the USA. These can also be used when hunting hares and can help avoid nasty accidents which could cripple our hawk for life. Photo courtesy of Francisco Solano.

3. Sew a small piece of leather on to the central space of the bottom part of the pattern (the one that would be the hawk's middle fore toe).

4. Place the chaps on to hawk using the same system as for Aylmeri anklets, first threading through one of the ends through the slit of the other end, then bringing them together using a metal grommet (again just like the Aylmeri anklets).

Please refer to the Appendix at the back of the book for some very simple hand-made chaps. You can also find much better and more elaborate designs made by Gary Brewer in his own website (see useful addresses) and in Gary Brewer's book "Buteos and Bushytails". Alternatively, you may also buy Gary's chaps online at Northwood's falconry store.

Creance

This is just a long string used when first training a bird, usually a bird being newly trained, that cannot be trusted to fly free yet or when accustoming a bird to flying in a new place. This is to prevent it from flying away unexpectedly and losing it. When flying the bird on a creance, it is best to find a smooth grass field or a place where the creance doesn't get tangled up. If this is not

171

The Red-tailed Hawk

possible, the bird can fly on a creance with a ring. Flying a bird on a creance during the training process avoids birds getting lost. We can easily make one of these at home with just some strong string, a lightweight dog swivel and snap (like the one used for glove leash) to be attached to the jesses and a heavy piece of wood at the other similar to a rolling pin (smaller) so it doesn't fly away with it. Twenty metres (22yrd) is more than enough and when our bird is comfortably flying to the fist or to the lure at this distance, we will then move on to flying free. Creances can also be bought in most falconry stores (also online) for a modest price.

Lures

The red-tailed hawk should be introduced to both types of lures; the traditional dragged or rabbit lure for hawks but also the swung (winged) lure typically used with falcons. In the United States, they are already ahead of us and have been doing this for years with red-tails, particularly when training them for squirrel hawking, as it develops more muscle mass and greater agility, teaching our bird to capture the lure in the air as if it were a squirrel jumping from tree to tree.

Here we can see a falconer using the swung lure or winged lure typically used with falcons which, with a few amendments, can also be used for red-tails. We can also see that the falconer is wearing a hawk vest and also carrying his telemetry system with him. Photo by Arjen Hartman.

4. Equipment

Hawking bag or hawk vests

These are quite useful for carrying food both during the initial stages of training and also out in the field, when hunting and there are many designs/styles. It is not in our interest to let our bird know where the food is and should only see it once it's on the glove or lure. We should wrap its food carefully in a cloth or aluminium foil, better than plastic, when carrying it inside our bags or vests. These bags usually have many pockets and are ideal for helping us carry all we need.

It is quite easy to find them nowadays, either through equipment makers, falconry clubs or leather shops. I personally prefer hawking vests as I find them more comfortable and also more aesthetically pleasing, and amongst my favourites are those made by Ben Long (see useful addresses at the back of the book).

Telemetry system

Telemetry is considered to be a revolutionary invention in the world of falconry and could be seen as the modern equivalent to the traditionally used falconry bells, although it is always recommendable to use both with almost all birds of prey.

Good news is that red-tailed hawks, once manned and properly trained do not usually get lost. If they suddenly do become stubborn or uncooperative one afternoon, we may be lucky enough to just be forced to grab a good book and sit at the bottom of a tree waiting for them to come down, or in the worst scenario, do a little tree climbing ourselves to get our hawk, but that will be about it.

Stories about red-tails getting lost are unusual, if not unheard of, as they tend to be very safe birds to fly due to their territorial nature; if they do travel and leave us for a while out on an extended hot pursuit of a bunny, sooner or later, they will always come back to what they consider their territory. Red-tails are amongst the safest and most loyal of all raptors, which also makes them ideal for beginners, however once we are ready to fly them free, we should always fly them both with falconry bells and a telemetry transmitter.

Introduction

The first telemetry units appeared during the 70s. They were originally based on gadgets devised to track and follow wild animals so they could be studied and were developed by naturalists with this purpose in mind.

Although telemetry is widely used nowadays, there are many falconers who are still unaware of all its advantages and fly their birds without telemetry, with an enormous risk of not being able to recover them if they get lost.

Generally speaking, telemetry should be used with ALL birds of prey when not creanced and not only just with longwings or falcons. Although there is a commonly accepted belief that these birds can travel greater distances and therefore more possibilities of getting lost, any bird of prey can suddenly get scared, actually get lost or simple decide to "leave" or let it self be carried away far from us. It is better to avoid this and a good telemetry set is a small price to pay in exchange for peace of mind.

What is it and how does it work?

The bird is fitted with a small round-shaped transmitter which sends out an electromagnetic signal of a certain frequency at regular intervals. The transmitters are powered by small batteries (which must be regularly checked although some brands may have a small battery status LED to warn us when the batteries are low or about to run out) and the weight varies depending on many factors, but some can weight as little as 3g (recommended weight for small raptors) which really will not get in the way or "weigh" our bird down at all.

Besides the transmitter, a telemetry set will require a receiver or receiving unit which the falconer should carry on him at all times when flying his birds free (this is also a good reason to try and find the smallest and lightest receiver but with good coverage as at times we may be running and it could be quite uncomfortable if we have to carry a heavy and large receiver). The receiver works by tuning in (just like radio units do) to the frequency of the transmitter. The receiver comes with a directional antenna (yagi) which captures the signal that our transmitter is sending. This signal will be stronger if we are pointing towards the transmitter (and if it is polarized in the same direction, that is if the transmitter's antenna is within the same range as the antenna of the receiver). In simple words, what this means is that we will be able to hear the signal louder the closer we get to our bird or vice-versa.

This means that if we want to know where our bird is if she's not directly in sight, we only have to switch on the receiver, do a 360 degree turn and start walking in the direction of the strongest or loudest signal. This will be where our bird is.

Frequencies are measured in megahertz and the most common used frequencies for falconry sets are: 151 MHz, 173 MHz, 216 MHz and 433 MHz. The use of frequencies is regulated by a commission in each country and can differ. Some frequencies can be freely used while for others it is necessary to pay to obtain a license for its use.

In North America, 216 MHz is the widely accepted frequency for telemetry units while the pan-European frequency is 169 MHz (although this is not used in many European countries such as Spain where many currently use 216 MHz and the UK where the main frequency is 173 MHz) but the frequency of the future to watch out for at least in Europe will be 433 MHz.

The differences in frequency and points which we should keep in mind when buying or telemetry set are basically how the signals are distributed, which legal frequency we are to use depending on where we live and the type of land where will fly. With lower the frequency (for example, 151 MHz), signals work around the land and may perhaps cover some obstacles such as hills and tree but will also be more likely to bounce back to

4. Equipment

us and lead us to a false point and not where our bird is. With higher frequencies such as (433 MHz) the signals go in a straight line and will be fainter if there are obstacles in the way (hills) but will be less likely to bounce back and will improve our overall efficiency in locating our bird.

Currently in Spain as most European countries (save the UK where 173 MHz is used) the recommended and legal frequency is 433 MHz which works well with flat or moderately flat hunting lands such as those used for flying longwings. If our flying land is hilly or has mountains in it, it is possible that the 433 MHz signal could be weaker or lost. In this case it would be better to use the American frequency of 216 MHz or the frequency used in the UK of 173 MHz.

There are however some disadvantages with regards to using the American frequency of 216 MHz in Spain and in Europe and that is that this is a reserved frequency to be only used for digital radio. It is not being widely used yet, but as this frequency is used more and more for radio it could create interferences and a lot of noise that could make very difficult (or almost impossible) to be able to receive and hear the signal of where our bird is.

Europe – flying with two transmitters or frequencies

This could be the perfect and safest solution but it is definitely not cheap and would require for two transmitters of two different frequencies (meaning you also have to buy a receiver for

Here we can see a female red-tailed hawk wearing a transmitter on a tail-mount, one of the best options for a hawk these days in addition to the back-pack option.
Photo by author.

each frequency), for example using 433 MHz and 216 MHz for Spain or perhaps if you live in the UK, acquiring a new receiver and transmitter in 433 MHz so you can use together with your unit at 173 Mhz. There are actually many falconers using nowadays not only one, but two transmitters on their bird (for example tail-mount and leg-mount or back-pack), as these can sometimes be lost or not work properly so it is not so uncommon, although many will use the same frequency for both. We could then use the 433 frequency for helping us to locate our bird in flatter but larger areas and the 216 (or 173) for emergency purposes as it can help us better in hilly country or where many obstacles are presents (trees, buildings, etc).

Obviously this means that you will need to buy two different telemetry sets with receivers and transmitters so double the expense, but some falconers justify this expense when flying very rare birds or when doing bird control as their income depends on it. In any case this option is quite expensive for most of us and also not too practical to be running around with two receivers on your back, unless they are of reduced size like for example Tinyloc's, whose small model is slightly larger than an out-of –fashion mobile phone which makes it very handy. Perhaps in the future, regardless of the frequency it would be good for telemetry brands to turn towards that direction and make their receivers smaller and think more about the falconer who has to carry them? That would be an idea…

Fitting the transmitter to your bird

Transmitters can be fitted several ways: neck-mounted (used in small raptors but not at all recommended), leg-mounted (easily attached with a bewit before flight and then taken off after), tail-mounted (attached to the main two deck-feathers) or as the latest craze suggests, using a back-pack mount which goes on the back of our bird and apparently seems to bother it less and not get in the way at all, though fitting it with some birds can be an issue, especially for beginners.

In conclusion, a telemetry system will always be the best investment you can make if you intent to seriously practice falconry and look after your bird's well-being and when buying one, we should never sacrifice quality and go for the cheapest price. This will be one of the most important tools you will have to help you if all goes wrong, remember that.

There are many different brands and types of telemetry systems available on the market currently with price tags to suit everyone. It would be too time-consuming and endless task to go into detail about them, for this again, I recommend visiting **www.yarakweb.com** and reading up on the subject there, however I can give you a few useful tips, regardless of your geographic location.

Budget-allowing, we could purchase some of the latest telemetry systems such as those made by the big telemetry brands. These are now quite small in size (some transmitters weighing only 3–9 grams which is 01.-03 oz), quite durable and water-resistant and with long-life batteries. The smaller the transmitter, the less it will be in the way of our red-tail and it will be less difficult for it to get caught on a branch or somewhere when hunting. To avoid this, we must also examine the different fitting types (see above about fitting a transmitter).

Though red-tails don't usually get lost, if we are going to invest in a telemetry system, we must make sure that we get one with the best possible range, as it will be our duty to watch over our red-tails and make sure that if they leave us while hunting a prey, we know where they are. This is our responsibility.

Transmitters and receiving units can now be bought of all shapes, colours and sizes practically and almost a la carte based on our tastes, but no matter how good they look, if these systems are not checked regularly to make sure they are working properly and that the batteries of the transmitter have not run out every time we fly our birds, they will be practically useless.

A final word of advice here is to watch out for rainy or stormy days, though most models are now water-resistant, our birds will be more likely to get scared or "lost" on these days and it will be very hard to find them under such circumstances. Not to mention that they will not only be scared but also wet and unable to fly great distances. If our transmitter gets wet and is not water-proof, we will loose the signal and with that our bird. Our red-tail will be wet and unable to fly as it should, perhaps even be in trouble and if we have used falconry bells as well, we may have a chance of finding it (or better said hearing it) but things could get quite ugly. If it lands on an electric power line with the transmitter (for this reason we should always avoid flying our birds in areas near to such electric cables, poles with transformers and pylons, just in case) it could get electrocuted and die.

I would firmly advise all falconers to stay away from areas with power lines. It is believed that 90% of captive bird deaths could be avoidable if falconers did this and stopped using conductible materials on birds, such as metal rings and telemetry devices. The answer obviously is not to stop using telemetry systems, as they can be vital for the recovery of our bird but to keep our eyes well open for any electric power lines and know the land we fly our birds on.

For more information on telemetry systems and also different brands, please visit our worldwide directory at **www.yarakweb.com**.

Transport

If we do not have a suitable hunting ground or hawking land near our home where we can fly our red-tail, we may need to travel further and will need to get her used to travelling in the car, preferably with a giant hood or transport box. These are best for transport, as raptors are in complete darkness and since they would be perched inside them, they would not fall or hurt themselves when driving over bumps, curves, etc. These giant hoods can be made at home without any problems to the measure we need for our car. They can also be bought online.

If we are driving alone in the car and have no other birds of prey with us, we could get our red-tail used to being perched on one of the seats (tethered), though if not properly trained and well-behaved this could be both dangerous for the driver (constant batings and flying around) and for the bird who could also be interested in what's outside the car or suddenly be scared of something.

As there are no legally approved seatbelts for red-tailed hawks in modern cars, we must think about safety first. If we are making very short trips, we may be able to drive with our red-tail perched but for longer trips or trips on highways, I would recommend a giant hood. It is without doubt the safest means of transport and raptors can spend quite some time in there without there being any danger of hurting themselves.

Bath pan

Although it may not look important, this is a very necessary piece of equipment for any bird and for red-tails, it must be of a good size. Red-tails do love water and love to bathe, though many will prefer a morning shower with the garden hose to a bath. Also, youngsters may be reluctant to bathe and may act more like a cat whose about to get wet than a raptor upon seeing a bath. Even so, it is important that every bird of prey have daily access to a garden with a bath pan where they can relax, bathe and most importantly, drink and preen their feathers with the sun.

There are many different types of bath pans available for falconry nowadays but one of the best and most inexpensive is to use a plastic water dish for plant-pots, like those sold in gardening and DIY shops.

I also must confess that with new birds, a garden hose will be very handy as most birds of prey will not bathe in captivity until they have been manned and trained. During its first year, Bandit rarely took a bath and I only saw him do so on a very hot day. Instead, he loved to be splashed with the water-hose, even in winter! As I have mentioned previously, red-tails really do love water in the wild (in fact, they could almost be mistaken for ducks!) where they can be seen regularly bathing and playing in the water and in falconry, even when chasing quarry, they will not hesitate to go after them in the water and even dive for them if we are not careful. Once trained they will never say no to a great shower (similar to that of heavy rainfall) and will even appear to treat us as their personal spa staff asking us to shower them as they spread their wings and turn their back to us.

4. Equipment

A weathering lawn with some unusual sun for the United Kingdom. Here we can see the birds perched and tethered, with their bath pans where they can take a dip and cool down. They must always have access to a shady area too, especially if we are not keeping an eye on them and it gets too hot. In the summer, it is more than recommended to regularly sprinkle our hawks with fresh water from a garden hose to freshen them up and relieve them from the heat. Falconry furniture by Ben Long. Photo by Arjen Hartman.

The Red-tailed Hawk

"Bandit" ready for take-off! Photo by Arturo Marzán Gil.

5. Basic Training

The arrival

Although it was quite some years ago now and I lived in Spain then, I still remember my trip to the UK to pick up the red-tails. The trip seemed to take forever; perhaps because I missed the country that had for so many years been my home, or more possibly because after what seemed an endless search, I had finally found (or so I thought) a breeder who would supply me with my dream red-tails. Its funny, but I even remember catching myself at times before the trip, fantasizing about what they would look like, especially the male red-tail that I would train and fly. As for the other two red-tails, a couple, they would be destined for breeding and there would not be too much handling involved.

Finding the red-tails I had in mind was a lot of hard work. I had already seen a few breeding centres in Spain and had almost actually bought one then, but I wasn't too convinced by what I saw. After reading up for years on my favourite raptor and its hunting possibilities, I was looking for something quite different. I had seen many red-tails, but none like the ones I was about to meet.

When I went to pick them up, I was amazed; they were huge. It was even a little scary to get close to them. They seemed so powerful and brave. They were beautiful too... much more than any other bird of prey I had ever seen. I was first introduced to the "parents", the couple that would become my breeding pair. The female seemed almost as big as a small eagle and what feet! Then, suddenly, my attention was drawn to a large transport box that began to move and shake. There was something inside that seemed to be kicking with all its strength and fighting to get out, perhaps a little dragon? We opened the door carefully and cautiously looked inside. The first thing we saw were a huge pair of powerful feet thrust forward, a sort of "red-tail greeting" in response to our "intrusion". After putting a gloved hand in, the breeder's face seemed to grow a little paler and pulled out an enormous and breathtaking mostly white red-tail. It was in that precise moment that he stole my heart... that is why I then decided to name him "Bandit". He would be my hunting companion and friend and was exactly how I had dreamed he would be.

The search

Once we have made our decision about trying out this beautiful and efficient bird in falconry and we have a solid base of some basic knowledge, we can then take this first step and search for the ideal hunting companion with whom we will share many moments of our lives.

If we already know exactly what type of Buteo jamaicensis we prefer, we can search for our hawk by contacting various breeders (those of you in the US will have to trap them, ask your sponsor about this and see the chapter on subspecies for geographic location). I have compiled a list of comprehensive breeders which I have included in the world wide directory of my website, please visit **www.yarakweb.com** to see it or add further listings.

No matter where we get our red-tail from, one important thing is that *she must always be parent reared* and not hand-reared as is the case with many red-tails in captivity. Please check this out when acquiring your bird and make sure that you get what you ask for. Hand-reared birds will be screamers, aggressive and very dangerous to handle, especially by inexperienced hands and we will be no match for a red-tail's grip when angry. They are incredibly powerful!

If we are in Europe and acquiring a captive-bred red-tail, note that these birds will also need full documentation, even if they are considered to be, along with Harris hawks, "exotic species" which means that they will not require a CITES permit (unless captured and imported from the wild) but will require all the other usual paperwork. If in doubt, please consult with local authorities.

Last but not least, I also recommend that your hawk be an "untouched" individual for best results. In other words, a chick of the current year that has not been manned or trained yet. It could otherwise have been trained wrongly and though red-tails are very forgiving and will tolerate quite a few mistakes (which also makes them ideal for beginners), some bad habits will be impossible to correct. This is also the reason why I always object to buying trained birds as many breeders offer, unless of course you really know the breeder well and how he trains them.

Initial stages of manning and training

This will be one of the most exciting and unforgettable moments that you will ever share with your hawk. It really will be very special, particularly when the wait for a bird has been a long one, to finally "meet" your hawk. We must make sure we have done everything possible to be as fully prepared for this as we can be. This is not the time to hesitate or make mistakes and we will need to know clearly in our mind every step of the way. Beginners will undoubtedly make mistakes, but we must discipline ourselves to avoid as many as possible. Though it is true that red-tails make excellent birds for beginners, I would like to emphasize here that I am referring to people who have been studying and reading up all they could on falconry, preparing for the arrival of their first bird, and not those who have a complete lack of knowledge on the subject. We cannot simply go out and "get" a hawk, call ourselves beginners, and then start asking falconers left and right what to do, or not to do, in falconry forums. This is irresponsible and luckily, illegal in many countries such as the United States and northern parts of Europe. To get the best out of this hawk and any other bird in falconry, we need to have some solid basic knowledge before we even think of getting our bird and calling ourselves beginners.

Once you have your bird, many of you will be in a hurry to train it as soon as possible and start hunting; however this is not the way to go. Some falconers even

5. Basic Training

go to such lengths as asking the breeder not to feed it the day before they pick it up. I think this is madness an I firmly believe that it is much better to wait to start dropping the weight until we have our hawk safely at home and have allowed her to rest for a few days, especially if we have had our hawk shipped from abroad. It is likely that with all the travelling, she will be suffering from a certain level of stress and will not have eaten much either. Consequently, she will not be at top weight when we get her. We must therefore, let her calm down and get all her strength back before we decide to begin manning and dropping her weight as that will be stressful in itself too. If we do not act with caution, we may be putting her life at risk as stressed birds are often weak and can easily fall ill. I have had many students come to me with apparently healthy birds who suddenly died just a few days after for no apparent reason, save the rush in trying to lower their weight to begin training.

Going back to our first day, we should have already bought or made a few hoods of different sizes and shapes for our red-tail (we will need to try several to make sure it is a proper fit, both shape and size). For this, it would be best to speak to the breeder if we are buying a captive bred bird as he will be able to guide us better regarding size. It will also be better, if it is possible, to go and pick our bird up ourselves instead of having it shipped to us. If our hawk gets a little too nervous and doesn't stop bating, if we bring a friend along, we will always be able to look after it better and more so if we bring it hooded and sitting on our glove (if we are not driving!). Having said this, we will also need to bring a couple of jesses with us, a large transport box, a proper falconry glove and have the rest of equipment ready for her arrival at home (see previous chapter). It is possible that she will be very nervous while travelling and often bating inside the transport box (which should have been made dark on the outside and padded on the inside to avoid ugly surprises). If this happens, we will have no other alternative to hood it and carry her on our glove while someone else drives to avoid the hawk hurting herself in any way.

Upon her arrival home (if the bird has been shipped) or when we go to pick her up from the breeder (or have just trapped her), we should check our red-tail out to make sure she is OK. We will examine the feathers for lice or damage and the base of the feet, not only for cuts and grazes but also for bumblefoot as red-tails being rather large do have a tendency towards this ailment. We must also then check its overall health and feel the breastbone for the first time. All of this of course, while our bird is hooded as captive bred or not, if she has been parent reared and untouched, she will be just like any red-tailed hawk born in the wild and quite scared of us. After we have checked that she is OK and armed her with jesses and anklets (see previous chapter), we must then leave her to rest in a calm place (if possible a little dark), tethered to a bow perch, and let her get to used to her new surroundings and having jesses for the first time. Obviously, she will not eat out of our fist that day and may not even be hungry, although if we leave some food in front of them and walk away, they will probably eat it when we are not there. It goes without saying that at this point, especially if we leave food, the red-tail should not be hooded. It will also be necessary to grease her jesses as she will attempt to bite through them now. If we like, we can also leave a nice radio station switched on for her to hear so she starts getting used to noise as well as human speech (which can accelerate manning) but I would recommend total peace and quiet this first day.

After one or two days, we will be able to begin the manning process which with red-tails, could well take a little longer than with other birds of prey, such as for example Harris hawks. This, however, can sometimes be accelerated a little if we use traditional medieval falconry methods such as "the wake or watch" to man them rather than the usual 15–20 minutes a day which modern falconers can spare and really is not sufficient with certain individuals. It may result in quite a few sleepless nights for both (see glossary for more details), so be sure to have plenty of time available for it (on vacation) or take turns with someone else, but the results will be noticeable in just a few days. This I have found it's the most effective method for me, as most red-tails will actually need a lot of manning, and the more we are with them, the better they perform as I later found out. We will hood our red-tail, secure her to our glove and take her to a partially dark and quiet room, after having weighed her first and felt the breastbone. We will then attempt to remove the hood very slowly and without looking directly into her eyes as she will feel intimidated and will take this as a threat, opening her wings and beak to us. Slowly, we will find a comfortable sofa or chair (do put newspapers all over the floor and on the sofa as she will propel her droppings and things could get very messy!) and I recommend a good quiet film too. The only lighting we will need here is a candle or fireplace and the light coming from the TV, so do get the room looking cosy! This will be the first "date" or moment that you will spend alone together and try to get to know one another. She will have to learn to trust us and get used to us a well as to sitting on a glove and will probably bate a few times; we will in turn, have to get used to her as well and to her weight on the glove.

At this point, it doesn't really matter much whether your red-tail is wild-caught or captive-bred; as long a she's a parent-reared eyass and has had practically no contact with man, they will be both almost just as wild and equally scared of man. This may mean that they will require a lot of "manning" and many hours of sitting "on the fist", something not really needed with imprints but believe me, an imprinted red-tail is a very dangerous bird of prey; not something beginners want to mess with. **The more time you spend with your red-tail on the fist, the more you will notice the results**. Red-tails can be quite heavy to carry on the arm after a few minutes, especially if they are constantly bating, so I recommend you find a good place to sit with an arm-rest so you can sit at a comfortable distance from your bird and keep a short leash to avoid more bating. Your arm will be quite sore the next day as you

5. Basic Training

will have to sit with her for at least a few hours now, but after a while she should not be bating too much and will tolerate us a little. Then, after some time, we can try to get her used to being stroked with a pigeon feather. Once she gets used to this, and seems more comfortable, it will be the moment to try and give her something to eat.

With our hawk tightly gripped by her jesses in our gloved hand (please make sure that she is also secured to your glove, as she could fly off), we must try to get her to eat a little on the fist. We must offer her a small piece of meat. The meat should be fresh and juicy (in this case with no feathers, bone or roughage material for casting as she will have a hood on), if possible a little bloody and previously warmed up in our hands so that it doesn't feel too cold to swallow. We will try to get our red-tail to take this piece of meat from our fingers, getting it close to her beak (but be careful!). A trick here is to spray the beak with a little water so and then get the meat close to her so she can get a taste when swallowing the water. This will encourage her to eat it. It usually doesn't take too long to get a red-tail to eat for the first time, even though they will have fat reserves and may not even be all that hungry, but they are gluttons and we will always be able to win them over with food. As Martin Hollinshead says in his book *The Complete Hare Hawk,* A bird of prey that has a full crop and will still do anything for another piece of meat is really the best bird that a falconer can have to work with and we can definitely use this to achieve results. The fact that red-tails love food and are natural gluttons means they will do anything, including to cooperate in training, just to get a piece of meat. This, together with a controlled weight reduction and plenty of time on the fist, is what will actually make the whole process of manning and training run smoothly.

If our hawk already feels comfortable with our presence, she will eat, if not, we will try this again later. We can also try to touch her feet with a piece of meat in our fingers. She will be used to our stroking but not to us touching her here and she will probably try to bite our fingers down below at her feet when we do this. Instead we will put the piece of meat against her beak so she can taste it and bite it. She may throw it out altogether or swallow it. If this works, we can repeat this a few more times and then gradually just attempt to give her pieces getting them close to her beak.

If our red-tail rejects food, we should not worry too much, even if this whole process goes on for more than 3 or 4 days, as that can also be normal, but in general they usually take 2 days to start eating. If she does eat, we will make a clicking sound every time she swallows a piece so that she begins to identify sound with food. However if she does not eat and takes a few days, or we are worried that our bird is too thin to start off with, an option is to leave some food in front of her when she is back in her hawk quarters, tethered to her perch, and unhooded so she can see the food. We will then leave her alone to eat at her leisure and she will most likely do so, as soon as we are not there. Once she has "settled down", we will be able to start gradually lowering her weight so that after manning we can begin with training.

The following day, we will carry out the same routine (always weighing our hawk before she is fed) spending a few hours with her, until she is comfortable with us and with sitting on the fist. Our next task will be to get her to eat properly on the first now, instead of getting the pieces of meat from our fingers, we will attempt to feed her a larger portion of fresh meat, for example half a quail or a pigeon breast. We will begin by offering her small pieces just as before and will then swap it with the big piece of meat. When she does "bite" it, we will slowly lower our hand towards our gloved hand so that she still pulls from this meat, letting her finish the entire piece on the glove. Until here, the training will be very similar if not identical to the training of most birds of prey used in hawking. Once she is eating well on the fist, having gradually increased the light in the room from a partially lit room to broad daylight and also the level of noise, without a hood, we will be able to start carrying her around on our fist while we take a walk outside and she is busy pulling at a tiring (for example a pigeon wing or chicken leg, etc.), getting her used to the movement of the fist, to people, to being outside, to the dog, horse (if we are going to use it), etc. It is important to note that red-tails work quite well both with dogs and horses as we have all seen in the movie "Ladyhawke", but they will have to be slowly introduced to them and this is best left for both experienced birds and falconers.

Working with Dogs and Horses

When introducing our red-tail to dogs, it will be best to not do this if we have a puppy and wait till the dog is a little older and has learnt the basic commands. I had a female Irish Setter puppy "Lisa" who really was quite a handful, not at all obedient and very, very loud. She really made my red-tail very nervous at this stage so I decided best to wait to introduce them to one another until she had grown up a little and calmed down. The best way to introduce them when the time is right is to first let them both eat together; the dog on the ground (tied up) and the bird secured on the fist. A key point here is that the dog we pretend to use for falconry should only eat dog food and must never actually be allowed to capture prey as such. If we do this, and our dog is not very obedient at one point, we could have a very unpleasant situation in our hands. We must note that the number one thing for a red-tail is her food and she will never tolerate anyone that comes between her and her prey. This is the most important thing to remember if we are to hunt with red-tails and dogs. Once they tolerate each other's presence, we must let them be together in the garden, tethered at a safe distance, to get used to one another. No matter what breed of dog we have, our dog will

5. Basic Training

need to be as obedient as possible, used to being on a leash and to cars, loud noises, etc. It must also be trained for flushing quarry in falconry and not a be a retrieving hunting dog as this will not work. If we do not have an obedient dog and he interferes with a red-tail's food, it could get seriously hurt. There are also certain breeds which are more appropriate than others for the type of land and quarry we wish to pursue. We should also keep this in mind and only move on to using dogs when we have some experience both in falconry and with our bird.

As for horses, this is definitely not for beginners (both hawk and falconer) and certainly not suitable for everyone. Though romantic as it may seem, it is in fact quite tricky as fact quite tricky as not all horses will tolerate a bird of prey. It can also be quite dangerous but I would like to run through a few main points to keep in mind to give readers a general overview. Indeed, most horses are easily scared of the most significant things not to mention flying or fluttering around them, so if we are to pursue horse-back hunting with the red-tail in falconry, we must select a horse with the right temperament. It must be an extremely obedient individual and not too nervous, as most of the time we will be riding with the reins in one hand. This horse must also be very agile and good at jumping, as at times the chase will be at great speeds and if we are flying a red-tail, there will surely be many obstacles in the way. The horse will also have to have a great deal of patience as it will have to wait for us when we get down to make in on our hawk. Last but not least, our ideal horse will also have to accept being close to our hawk. The best candidate for horse-back hunting with the red-tail will most likely be a horse that has grown up in a stable where there were pigeons or doves nesting. It will be used to the fluttering noise of their wings and also to seeing them fly, possibly around it, so will be more inclined to accept any bird that flies towards it or around it. If we already have such a horse, then we really have a great advantage as at least the irrational fear of birds will

The author's father training a horse, "Wellington", and working with dogs. Photo courtesy of Antonio J. Candil.

not be another worry to add to the list.

The best way to introduce our horse to our red-tail will be similar to that used with dogs, first getting them used to the movements of one another, starting with getting them to eat together. We will also have to get the red-tail to be on the fist while the horse is on the move, something that is not always easy for both bird and falconer as I also found out. The best way is again, just like we did when she was getting used to our walking while sitting on the fist, is to use a tiring and get her used to the movements of the horse. Little by little increasing the speed and avoiding her to lose her concentration from the chase. For this reason I do not recommend to start hunting on horse back with inexperienced birds but rather birds that have hunted for a few seasons at least as their complete focus will be on the hunt and will ignore everything else and concentrate on their prey. We will also have to get both red-tail and horse to get used to the red-tail flying in to the fist from a distance or from the horse to the field. This is also quite tricky and must not be attempted at all until both horse and bird really have shown that they really get on together as otherwise a big accident could happen and we could really get hurt. For this, we will need a garnished lure and will need to throw it out in front of the horse but not too close. Just a sufficient distance for it to be near enough without alarming both bird and horse. Little by little, when they seem used to this, we will be able to decrease the distance and throw the lure a little nearer until we can try calling our red-tail from a post to where we are. We will then also try to call her to the fist but with us standing just a few metres in front of the horse, and also decrease this distance little by little till we are almost next to the horse. Next, once used to this, we will then do the same but standing on a stool or something that makes us look taller (a bench, nearby fountain) and next to the horse but on the other side, so that our hawk has to fly over the horse to get to our fist. In this way, we will get the horse used to our bird flying in to us and in the event that it gets scared and tries to run off or throw us off, we will not be sitting on top and will save us from a nasty accident. This technique has worked well with many falconers, however before we put anything into practice, we must remember that not all horses will be good candidates for falconry and some birds may not tolerate them or they just may not be compatible.

In any case, we should note that we will not be able to force such cooperative relationships on animals and even when things seem to go smoothly, as they did in my case, unexpected things can suddenly happen while horse-riding in the silliest ways as most horses seemed to be quite scared the most insignificant things. In the summer of 2004, I dislocated my left shoulder and had multiple fractures on my left arm which had to be operated as a result of going for a routine ride with my usual horse. After riding for the last two hours, I was making my way home when suddenly my horse noticed a small snake and just got completely out of control. It decided to give me a small taste of what a rodeo could be like before finally managing to throw me off and running

5. Basic Training

away back to the stables. I fell hard on the rocks and my shoulder was instantly out of place; the rest is quite an unpleasant story. Luckily I did not have my bird with me on that day or things could have been even uglier. It took me more than two years of daily physiotherapy sessions and more than 8 hours of exercise a day to recover and to this day, almost four years after my accident, I still cannot use my left arm fully or carry any weight on it at all. This is what can sometimes happen if you decide to ride a horse, let alone combine it with falconry. No matter how many years you have been riding, as my father once said, "if you really ride, often sooner or later you will fall and have accidents. Those who claim they do not fall are those who actually don't ride, at least not on a regular basis". He has been riding professionally for more than 40 years, even doing show-jumping and to be honest, he's had some of the worst accidents I have ever seen.

Needless to say, we will also have to get our bird used to riding on the glove and various speeds and this can also be quite tricky, that's why again, this is not recommended for beginners: whether they are falconer or hawk.

Going back to general training; when our hawk is finally comfortable with us and her surroundings, having spent many hours on the first, it will then be the right moment to go on to the next training phase: jumps to the fist. For this, we will have to secure our red-tail to our glove. Then, with our hawk on our fist, and as she eats from a good piece of meat on it, when we see her really get stuck in, we will then put her on some nearby perch (it could even be the back of a chair) and take our hand away a short distance from it. Just enough so that she doesn't reach the meat but clearly sees it almost within reach. In this way, sooner or later she will jump to our first for it. Here is where we will need our patience again and where many falconers get discouraged as other birds, like Harris hawks, may perhaps perform quicker at this stage, and although not all red-tails are the same, this first jump will generally be really hard work for our hawk.

It is just a matter of "tempting" her, giving her a small piece of meat so that she tries it and so that she is still hungry enough to jump for the rest. She should be hungry enough to want to jump for the meat but not so hungry that she is too weak and just sees it as too much of an effort and waste of energy. This is also why we should never lower the weight of a hawk in one go or as fast as possible, as she will turn aggressive. There is a key word that we must have "engraved" in our thoughts when training red-tails: *patience*. We will be able to achieve wonderful things with them but only if we carry out a proper training and for this, we will require enormous patience at times. If your hawk looks at the glove but does not even attempt to jump, not even to stretch out a leg for it as many red-tails do (and its quite funny to see), then her weight is too high and we need to drop it down a little more and give it another try later. My red-tail Bandit was quite

189

frustrating in the beginning as he really seemed to respond well to manning and his manners were exquisite and outstanding, I really couldn't complain, but he was so stubborn that it was impossible to get him to jump. I really had to drop his weight down several times for this. He would then make a vague attempt, try to stretch his leg out, would foot my glove, and would then look at me and shout at me as if he was saying "what do you want from me?". He really didn't seem to understand what I required from him and seemed to be more like a plane with its engine roaring, about to take off, than a red-tailed hawk. He would do anything but jump. Each day he seemed to be closer to it, then at the end of the day I would go home empty handed, I almost gave up… until finally, one snowy afternoon, while we were both shivering outside, he decided to jump.

Advanced training

The first jump is always the hardest. Once they have accomplished this, the following jumps will be much easier. With Bandit, after his first jump, everything went really smoothly and seemed too good to be true. He was waiting anxiously for me everyday and was more than ready to jump to the first, it was as if it was his way of greeting me now, showing me that he was a keen student and willing to learn so much more. I gradually increased the distance at which he jumped, still secured to my glove, now using the whole length of the leash as a measure and in a few days, Bandit was ready to move on to flying outside. The time had come to introduce him to the lure and slowly show him the world he could have at his feet… and then suddenly, everything came to a halt. Bandit was motionless and seemed to ignore me, admiring the beauty of life, watching the world go by as if time stood still for him. I really don't know what was going through his mind then, he seemed to be interested in everything, from butterflies to the flowers gently swaying in the breeze, or nothing in particular. Bandit would just sit there, daydreaming the afternoon away when it was training time. He was at the correct flying weight, was comfortable with my presence (even appeared to enjoy it) and we would see each other a few times a day, every day. So what was going on? The answer was simple: he was panic-stricken!

Once again I had to dig deep down within myself for that little bit more of patience that we all have hidden somewhere (very hidden in my case!). That same day, after several more attempts on my part, Bandit and I decided it was better to take some time out and see how we could go on from here. That's the thing with red-tails, they really make you think about every move you make and there is always a reason behind everything they do. I thought I had a keen and willing student, but perhaps I had understood wrong. In fact, I didn't know it then but I was to become a student myself, and he would be the master who would discover things before my eyes. That is what he was trying to tell me I guess, now I see it, but then, I just thought Bandit was being stubborn. When things don't work out the way we want them to it's too easy to put the blame on others and in this case, I almost fell victim to the prejudice on this hawk. I am not a quitter, so decided to lower his weight a little more and get him more used to being outside by taking him for more walks or simply sitting outside with him. After all, he had been chamber-raised; he didn't know what trees or clouds where or anything else for that matter, so maybe he was just a little overwhelmed with it all?

During this training phase of jumps to the fist, I would try to fly Bandit every other day, feeding him up well on the days he did good and giving him little or no food on the day after to maintain his weight. I have found this technique works quite well with most birds of prey and seems to really "fix" things in their head. I also decided to put Bandit outside all day long, in a weathering enclosure where he could relax and sunbathe all day, from early in the morning till the time he would fly in the afternoon. I thought this would also help to get him used to things. I would not leave him alone all day, but would also go see him once in a while, sometimes bringing him a little tidbit of food. After a few minutes, I would go away again to come back after a while, maybe spending an hour or so reading a book next to him. After a few days, he had lost only a little more weight, almost an insignificant amount, but seemed calmer outside and now had a certain exploring look in his face. I decided to try him again, this time at a shorter distance than what he was used to indoors and, to my surprise, he flew to my fist like he never did before. The first time I whistled, he actually took a little height even and before I noticed had landed on my glove and already eaten his reward. I tried it again once more, and he left me open-mouthed; he really had such a beautiful way of flying...

I gradually increased the distance, and then still on a creance, kept flying him every other day without too many problems, though once in a while, Bandit spotted a mouse which seemed to attract his interest more than my glove, but thankfully soon came to his senses and abandoned what could have been an interesting career in mouse-chasing. He was just curious that's all. He no longer seemed scared, though he did seem quite irritated with the neighbour's dog that wouldn't stop barking at him. I still think that to this day, Bandit still has it in for that dog (and not any other dog, as he was fine with Lisa, my Irish Setter). More often than I'd like to admit he seemed to suddenly wander by mistake into next door's garden to have a "friendly" chat with the dog...

I then lived outside Madrid, in a mountain region near to the ski resorts, and the weather was harsh and sometimes, it was just impossible to fly him. It would be hailing or there would be an unexpected snow storm and while waiting for it to go away, it would just get too dark. Those days I would exercise Bandit just the same but indoors, doing vertical jumps to the fist or jump-ups. In this way, he would still get a good workout, build muscle mass and not forget his training. He also started gaining weight, though he seemed stronger than ever and not at all fat. In Chapter 6, Jim Gwiazdzinski made a point in his article about weight gain which I'd like to recall now and the mistake that most people make with this when training birds. I have to agree with him and to be honest, its common sense. We all know that if we go to the gym and start working out, we may lose body mass and go down a clothes size or two, but when we really start building muscle up, we may even weight a little more. How is this possible? Well, simple. Muscle weights much more than fat, though it takes much less room too. We must always keep this in mind when training birds of prey too. All too often beginners are doing well with their birds and suddenly they undo their training by lowering the bird's weight. Although their hawk flies well, they just think it's put on weight and is just too fat! This is such a mistake! By doing this, they will be weakening their bird, and of course, the next time round, the bird will be at a lower weight (the proper weight for their standards) but their hawk will now not fly… she will be too weak to do so. So, they will lower the weight again, but with no results and in the end, the bird will either get sick or die. If we are flying our bird regularly, she is performing well and we can feel that she is well muscled and not fat, the increase in weight will be normal so please do not rush to lower her weight. If we do so, we will be putting a brake on the training and will have to start all over again, or worse.

Lure training

The red-tail is now coming to the fist as soon as we call her with the creance at a distance of 20–30 metres (or 66–98 ft, the maximum distance a bird of prey should be flown with a creance as she could get tangled up or caught up in something). Now is the moment to introduce her to the lure.

Up until a few years ago, the lure used with hawks was the rabbit lure or dragged lure, although some more adventurous falconers had even used a remote controlled car dressed up as a rabbit (I think the correct term for this nowadays would be "robo-bait") which can certainly be great fun and give your hawk a great workout. For red-tails, we can keep using the same lure to help entering her into hunting rabbit and hare, although, you will notice that with most red-tails, this is a natural instinct in them and from the first day they see something "furry", will be naturally attracted to it.

The fun doesn't stop here though. There is no reason why we cannot also introduce the red-tail to a feathered or winged lure, just like the ones used with falcons, but with perhaps a little more weight in it. This will allow us to use the technique of soar hawking with our hawk and also exercise her more to develop stronger muscles and a greater agility. This system had been used for years by Gary Brewer in the US amongst others to train red-tails for squirrel hawking and can also become part of our everyday training. It will also be of great use if we wish to pursue duck hunting as well as other feathered quarry.

Both dragged and swung lures are also sold by many falconry equipment makers but we can also make them ourselves at home. For the rabbit lure, we will need a rabbit skin/fur, and a long leash. We will

5. Basic Training

sew the skin to make a little long bag and fill it up with sand. We should have a few layers of leather on this before we add the rabbit skin and make sure we sew it tight, to stop the red-tail from breaking into it and swallowing it (approximate weight 500g or 17.6 oz). For the swung lure typically used with falcons, it will be the same type and size of that used for falcons, but as I mentioned, just a little bit heavier (about 300g / 10.6 oz), however we have to be careful how we use this lure as if thrown the wrong way, it could injure our hawk.

Using a lure is quite simple, but will require the help of another person, especially in the beginning stages. While we hold our red-tail on the fist by the jesses, a friend will tug and pull the leash of the rabbit lure, eventually dragging it. We will have tied a piece of meat to it, preferably on the part of the lure that is supposed to be the head so they learn to go for this first (though red-tails again, usually do this out of instinct as I saw with my own red-tail). If it doesn't always work out don't worry, should your red-tail grab somewhere else, male or female, it will still have a good chance of succeeding due to its enormous gripping and squeezing power, but this is where we want them to aim for. Previously, we will have familiarised the red-tail with the lure in the following way: calling her to the first, as we have been doing so far but holding in our fist a garnished lure. As soon as the red-tail lands on the fist and begins to eat, we will use our right hand to carefully take it from her and throw it on the ground at a short distance from us (1 metre approximately or 3.3 ft) and observe that the red-tail will now jump from the fist to the lure. While she is eating on it on the ground, we will again take it away from her (you sometimes have to be quite quick with this!) and let our friend take the red-tail on his fist about 2–3 metres (6.6–10 ft) away from us. Immediately we will throw out the lure in front of them and drag it a little. The red-tail will jump on it, taking it for a real prey and we will then let her finish eating on it completely. As she is eating though, we will slowly walk towards her, speaking to her to get her used to our presence (this will make it easier for making in on her) and attempt to collect the lure from her. The first time we ever come with this situation face to face, as well as with her first kill, we will finally realise that red-tails are much more similar to eagles than buteos. The key here is to let her "calm down" as she will be excited and not able to control herself. Red-tails really live for the hunt and they get into such a state of excitement that it can at times be quite dangerous for the falconer as the are so focused on the chase and prey that they could even hurt us without meaning to if we get in the way. A good trick is to keep walking around her, giving her tidbit of meat while we also talk to her. This will help to calm her down and after a while, we can even sit next to her and secure her to the glove while she is eating. We should let her eat on the lure till she has a full crop, though we won't be able to fly her the next day. Again, we can try flying her every other day like this until she gets used to it and is comfortable with the lure.

The Red-tailed Hawk

For the swung lure, we will need leather, a pair of pigeon wings (or wings of any bird we are thinking of entering her into as a potential quarry), and a leash. We will fill it up with sand and wool to give it a little (maximum 300 g/10.6 oz) and avoid hurting our hawk's feet in flight if it is thrust too hard. We will introduce the hawk to this lure in the same way as for the rabbit lure, but after 2 or 3 lure sessions, we will begin to throw it up in the air so that she gets used to catching it in flight. Be careful how you throw the lure or you could actually hit and injure your hawk. If this does happen, even if she has not been seriously hurt, it could still take her a long time, if ever, to recover from the shock.

The way in which we collect our hawk from the lure is also extremely important here as if we don't do it right from the first day, she will start picking up bad habits. Amongst these, that of "carrying", as well as footing us or being very difficult to pick up from the glove. Once we get her used to our presence (see above), it will not be a problem to get close to her. As soon as she has finished eating, we will pick her up from the lure as if she were a falcon. With meat on our gloved hand, we must put our glove in between the red-tail and the lure. Once she has grabbed this with one foot, we must slowly but firmly lift our hand upwards so that she loses her balance and has no other choice but to put her other foot now also on the gloved hand. At the same time, we will have picked the lure up with our right hand and quickly put it out of sight in our bag, while she is now busy with the meat on the glove. Another way of picking our red-tail up, especially if she's being rather stubborn and not wanting to let go, is to sit on the ground and swiftly take the lure out from her, underneath our legs while we keep her busy with meat in our gloved hand. Again, we will put this lure away quickly in our hawking bag. We must note for the future that whether we are flying our red-tail in training or she's out hunting with us, if she has failed a slip, we must always end the day by calling her to the lure, or also to the fist so that she doesn't forget them. These will be used as methods to retrieve our hawk when she refuses to cooperate so the more we use them, the more alternatives we will have.

Flying free

Finally the day where everything will be put to the test has come! You will probably be nervous and a little anxious, but overall possibly overwhelmed with excitement. Though birds are not usually lost the first day they fly free, there are some precautions that you will need to take. We should only move on to this final step when our red-tail is flying and jumping to fist as soon as she is called, as well as coming to the lure on the first attempt and with no hesitation in distances of 20 to 30 metres (866–898 ft).

By now, we will probably have swapped the training jesses (traditional jesses) for the Aylmeri system with anklets, with interchangeable jesses for hunting and for being tethered (see Chapter 4). This is really essential at this point, as you will have noticed that red-tails are quite "rough" when they are hunting and will

5. Basic Training

crash through any obstacle when chasing a prey. If a red-tail were to wear traditional jesses, not only would they get in the way, but sooner or later they would result in a dangerous outcome: our hawk could get caught up in a branch or somewhere, making it impossible to get her back (if she gets caught at the top of a tree, we would have to climb up for her), and could cause her death.

When we finally choose to fly our hawk loose for the very first time, we should try to choose a day where the weather is also good and never fly a bird on a stormy day. Not only could she get scared (most birds that get lost do so with bad weather) and not know how to come back, but if she doesn't yet know how to hunt (as could be the case with captive bred birds who have never have hunted yet), she could die of starvation. She could also get electrocuted. Another thing to keep in mind is to fly her in a place she knows, so that she feels comfortable and there is less risk of losing her. If we fly her loose for the first time, she will not notice that she is not creanced and will generally not make an attempt to fly away. Even so, I do recommend that you avoid noisy areas, any distraction and anything new that could scare her, and when we do fly her, it will be a shorter distance today. The aim here is that the red-tail should not be aware of this "big" change.

Although I have covered the use of falconry bells in the previous chapter, again I would like to state how important these are. Your hawk should always be fitted with falconry bells, even when we fly her indoors the first few days and during the whole training process and flights on a creance. When we go on to fly our hawk loose, eve if we use a telemetry system, if she is fitted with falconry bells, this will be an extra help for finding her or knowing where she is when we can't see her. If the telemetry system doesn't work or the battery of the transmitter has run out but she is near, we will be able to locate her by hearing her. If we do manage to loose our red-tail, the most important thing besides notifying the relevant clubs and falconers in the area that she has gone missing will be to look for her in her usual flying place. Red-tails usually don't stray too far and as they are quite territorial, they will always come back to what they consider "their territory". For this reason, we should look in the same area where we lost her, daily, especially before sunrise and sunset (as during the night she could seek refuge in a tree) and call her with the lure or to the fist. If a few days have gone by, the battery of the transmitter could have run out, but if she is fitted with falconry bells, it could help us find her if she is near or has now come back in the area.

With regards to flying areas or hunting grounds (be sure to also seek permission when you fly if needed on private land or hunting grounds), if we always fly our red-tail near our home and can walk to this area without having to get in the car, we may not need to think about hunting elsewhere. However, we may wish to travel further at times, or when we go hunting with friends and will need to use the car to get to our destination. We will also need to make sure that our hawk is comfortable in the car and has been previously introduced to it. Red-tails can be transported in many ways (see previous chapter on equipment). If we are going hunting together with a friend, the easiest thing will be to have our friend hold our red-tail firmly on the fist while we drive, though it will probably be more convenient to use a perch in the back part of the car or a giant hood. To get a hawk used to the car, we must first feed her on the perch we intent to use in the back of the car or perched inside the giant hood with the car parked. When she is used to eating on this perch, little by little, with the help of a tiring, we can get her used to movement, taking her for small drives. We will do the same procedure, but using a friend, to get our hawk used to the car, should we wish to pursue car-hawking (please check to see if this is legal in your area). If she reacts well, this could be a good way of transporting her although on days when they are right on hunting weight, she could be a little nervous. Needless to say, if we have other birds in the car, we must keep them separate. A giant hood or transport box will be useful in such cases and also if we never manage to get our hawk used to the car. She will always be much calmer in these, especially if she's in the dark.

Flying techniques: soar hawking and hawking

Although red-tailed hawks have been used mainly for hawking, unlike most birds of prey used for falconry, they can use a wide combination of hunting techniques, including soar-hawking as falcons do. This is what makes them one of the most versatile hunting hawks available to falconers and in my opinion a bird for the future of falconry.

Soar hawking

A technique which red-tails naturally use in the wild, this is simply to fly your hawk like a falcon. What I am referring to here is to actually fly your red-tail allowing her to mount and soar like a falcon, letting her perform a spectacular stoop as she spots her prey or is flushed from cover while she later strikes or binds to it. Not all hawks will react well to this (although some Harris hawks have apparently also been flown in this way), but red-tails will use it as an advantage and they naturally excel in this. Traditionally, hawks are flown from the fist and this technique will be extremely useful still for many slips but we should also keep in mind all of the natural hunting techniques employed by red-tailed hawks and their subspecies. At times, another technique may be more appropriate for a certain quarry or situation. This is where it will come in handy to teach our bird all we can and the more means we give her, the better she will perform.

Although there are many different subspecies of red-tails, as we have seen in Chapter 2, red-tails will never be the fastest of all birds of prey. However, most red-tails, if in optimum health conditions and regularly flown, will be incredibly fast and strong, bolting from the fist almost as fast as any goshawk or at times, even better. Moreover, red-tails possess the added advantage that they will not tire out after an initial sprint but rather keep hot on the tail of their prey until they eventually capture it. On the other hand, this way of flying may not be appropriate for all situations and our best option at times may be to let our red-tail fly free and let her decide when it's the right time as soon as she have spotted a suitable prey or we have flushed it out from cover. Being at a certain height will always mean an extra advantage for both spotting and attacking prey and if we always fly them from the fist, we could in fact be limiting their performance out on the field. If we really want to get the most out of this wonderful hawk and all she has to offer falconers, we should also explore the possibilities of soar hawking.

Soar hawking is just an example of one of the many advantages that flying a red-tail supposes as there are not so many birds of prey who can boast of such versatility. To see a red-tail soar is something that I cannot find words to describe. One has to see it and live it to really believe it and not many birds of prey can rival with a red-tail's dominion of the skies. This technique has been used by American falconers with red-tails for many years now and can be seen in many of the videos of the yearly sky trials or meets, such as those for example of the California Hawking Club. Though a natural technique of red-tails, year after year it never ceases to amaze spectators and is really efficient. Red-tails are not only capable of mounting up and waiting on for the falconer to spot a prey or flush it from cover, working well with dogs, but they are more than capable of performing beautiful and impressive stoops of more than 200 Km/h.

These buteos, are known by many as the "wind masters" as they are experts in soar hawking and should we decide to introduce them to this hunting technique, we will be surprised at how easy it is as they have a natural tendency towards it. Red-tails, will use thermal currents and the wind deflected upwards (slope soaring) like no other, soaring to heights out of view from us without any difficulty. This could also perhaps turn into a cause for concern, as if we do not take care we could actually lose them (many red-tails use thermal currents to ascend and then suddenly "disappear". Our fear for losing our red-tail could at times become irrational as a result of this, however, if we get our red-tail used to flying in a certain area all the time, she will recognize it as her own territory and will never fly too far from it or go astray.

To soar or not to soar; that is the question. If we are considering this flying technique, then this is how our red-tail should make her first slips. If we complete her training and she is used to flying from tree to tree or from the first (traditional hawking), it will be quite awkward to then introduce her to soaring. Most likely, we will end up with a red-tail that just perches somewhere as soon as we let her free, waiting for us to flush the prey from cover instead of soaring up in the sky.

To begin using this technique, the best moment is when flying conditions for this technique are optimum (when we find an ideal open space, with not too many bushes but plenty of quarry for example and the weather is good). As a falconer, you will be responsible for constantly flushing or providing quarry to your red-tail so that she doesn't get discouraged and keeps building up her confidence. It is really important for your red-tail to make a few kills during these initial flights as otherwise, she could lose faith in herself altogether. The rest of the training will be exactly the same as that for a falcon. This means that we must let our red-tail take off from our first when the time is right, upwind, and wait to see her reaction, calling her back before she has a chance to perch anywhere. Little by little, we will be increasing the time that our red-tail stays up in the sky until she begins to actually wait in above us and once she does, then call her back with a whistle. We will also need to teach her to soar higher and higher; here is where baggies (where legal) can be quite useful. For red-tails, the best flying day will be a day with a slight breeze and with many thermal currents, as differently from falcons and due to their larger size, they will need a good deal of strength to get themselves up and away from the ground. A red-tailed hawk will only be able to soar if she finds a thermal current or wind deflected upwards from a hill to help her ascend, as if she has to use all her strength to climb up it simply will not do it and just perch somewhere. This in turn will only contribute to more frustration and help to discourage our red-tail even more from ever attempting to soar again. It is essential to also remember that red-tails that have been introduced to the technique of soar hawking must always be collected with a lure if no prey is to be seen, in the same way that we would do with a falcon.

Hawking

Flying from fist

With regards to traditional hawking, there are two main techniques used. In Spain, like many other European countries, soar hawking is not used with hawks and these are generally only flown from the fist at quarry. I firmly believe that it is in our best interests to introduce red-tailed hawks to all techniques available for getting the most out of them as for example in certain situations, they will under perform or simply be limited because of our own fault.

For these types of flights, which involve our red-tail bolting from the fist towards a prey, we will need both strength and speed and to have an appropriate slip setting. Just remember though, that even though a red-tail will never match up to the initial sprint of a gos, if it is well muscled and fit, will never match up to the sprint of a gos, although if well muscled and fit, will in fact make it to the end of any pursuit and eventually even catch up to its prey, though flying from a height will always play to its advantage and maximise her chances for success.

Following -on

Some people will actually confuse this technique with soar hawking when in fact, it is just a variant of traditional hawking techniques which were typically only used with accipiters; our red-tail will still be very much flying at relatively low height or close to the ground as well as close to the falconer. This technique has become more popular in recent years in mainland Europe though hawks were flown in this way for quite some time in the United Kingdom. Known also by the name of "free-following" or flying your hawk from trees, it is curious to note that some falconers in Spain even call this "English-style of flying".

This is a very simple and easy technique and will require only a few lessons although some broadwings can be a little stubborn and may prefer to still-hunt instead. The first step is to take our hawk to a suitable area (calm and with many trees). We will first place our hawk on a low tree branch, not too far away from us, and then walk a few steps away and call her to the raised glove (with food) using a whistle. We will then let her feed on a good piece of meat (preferably her favourite type) as this will please our hawk twice as much; not only will she learn to fly to the fist for a bit of meat, but she will be getting a treat too.

We will again, place her on a tree branch and walk a little further now, turning around and then call her again with food and the whistle. We can also call out her name. We will do this a few more times. Then, the next time we call her, if we wish her to follow us, but not land on our glove (just follow us from tree to tree, or perch to perch as we are walking), we will hide our glove as she approaches us when we call her again and make sure we are next to a tree. As she will then have nowhere to land, she will perch on the nearest place to us, directly above us (a tree, street light, a tree or even a road sign). We will now throw her towards the tree branch so she perches on it and continue to call her alternating the food on the fist and using the whistle until she gets used to it and starts following us when we walk. We will always end the training session calling her to the fist and letting her feed up there. Some hawks may not readily accept this new exercise and if reluctant, instead of throwing them towards a tree, we can just gently perch them on a branch and slowly walk a little distance away. Normally, the reaction of our bird should be that of following us (this technique can also be used with red-tails or hawks that get a little stubborn when out on the field and refuse to come down from a tree when called) and instead of getting out the glove, we will let them perch somewhere near us. Then, we will proceed to call her to the fist with a little food. Afterwards, we will throw her towards the trees and walk a little so that she follows once again, calling her to the glove once in a while, alternative flights with food and without food so that she never knows when she is actually going to get food and just comes to the glove when she is called because she wants to come. Once our red-tail is used to this technique and follows us without a problem, we should try to set up a slip using bagged quarry (where legal) so that she fully understands the advantage of this technique and what it means to fly from a certain height (another extra advantage) with regard to hunting prey. This technique will not only give us hours of pleasure as we take a nice walk and considerably exercise our hawk at the same time, but will be another useful option to be used in certain hunting situations by our hawk.

The Red-tailed Hawk

Car Hawking

Recent years have seen the rise of yet another popular but controversial hawking technique which now seems to have taken over in falconry forums and videos mainly in the US but also in some parts of Europe (such as for example, Spain) where legal. The practice had been used for years by falconers doing professional bird control services at airports and other sites and I myself learned about it in this way. A few urban falconers also began to pick it up for flying their American Kestrels and small accipiters and now, It seems widely practiced by many, although an equal number are also against it. This technique involves driving a car along country roads or places with abundant quarry, spotting a victim and releasing the hawk out of the window so that it can catch the quarry by surprise. In other words, it is simply flying a hawk at a given quarry, from a vehicle.

Traditional falconers frown at this technique, many calling it "ugly falconry" or arguing that in fact, it isn't even falconry. As Matt Mullenix goes on to say "car hawking lacks almost every inherent virtue of sport, save that it can be a lot of fun and provide bagfuls of food for a hawk". One only has to go online to both our own website (www.yarakweb.com) or Youtube and search for elaborate falconry videos with overpowering sound-tracks to find a large number related to car hawking. I have myself tried this technique when I was involved in bird control and seen its effects. The prey that is being hunted, does not see the hawk in the car; and in fact, acts pretty much like an unsuspecting sitting duck. This technique can therefore provide bags full of game for you and your hawk, and when there is a shortage of game in breeding season it can be useful. Aside this and for pest control, I have to state that no matter how much fun it can be, it is definitely unfair and far from belonging to the definition of what falconry as a noble sport should be.

The practice is also heavily regulated in most countries though at times the legal boundaries do not seem so clear. For example, in the US, apparently no game species for which there are legal seasons may be taken from a vehicle. The law actually even goes on to discourage the practice of shooting deer along the roadside along with other species. These laws are not specific to gun hunters, so they can be expected to apply to falconers as well. However, one can find a legal "vacuum" here and work around the law through the unprotected pest species such as the following: starlings (*Sturnus vulgaris*), house sparrows (*Passer domesticus*) and pigeons (*Columbia livia*). In general, the hunting of these species has not been regulated yet as such (save certain exceptions such as bird sanctuaries). Apparently, provided falconers pursue these quarries on private property (i.e., from private roads), there seems to be no restriction to hunting quarry with a hawk from a vehicle. In the UK, however, any type of hunting from a vehicle is completely illegal as Section 5 (1)(e) of the W&CA 1981 states: "[i] If any person uses a mechanically propelled vehicle in pursuit of a wild bird for the purpose of killing or taking that bird, they shall be guilty of an offence." Unlike legislation in the US, the act does not specify where it is illegal whether it be a public road, in a field or on a country track, and therefore applies to anywhere within the UK.

The last step: entering into quarry

This last step can vary amongst falconers and will also depend on your country's (or federal) legislation on falconry. The following pages explain how I entered my hawk using bagged quarry (legal in Spain).

As soon as our red-tail is flying well to the lure and to the fist, we will move on to entering her into quarry. Before we start setting up slips with bagged quarry left and right, we should have clearly decided which is the quarry we would like to pursue. Logically, we should only attempt to pursue available quarry in the area, however, for red-tails or hawks, this does not have to be limited to "just rabbits". Red-tails possess an enormous potential for falconry and we should make use of this. Being able to use diverse hunting techniques will also assure that we have quite a wide range of prey available and not be limited to just rabbit and hare, as is the case with most hawks.

First, we must check out the area where we intend to fly to choose the best zones and see our possibilities regarding quarry. In my case, I had plenty of rabbit, duck and goose where I lived, so I decided to make these my main quarry. Once you know your prey, we should then decide the hunting techniques we will want to use. If we want to fly our hawk through soar hawking as well as traditional hawking, this will have to come first as otherwise, we will never manage to get her to soar and she will most likely find a post to perch as soon as she leaves our fist.

After deciding which quarry we would like to pursue, we will introduce our red-tail to them following a logical order. For example, here in Spain, we are not allowed to hunt squirrels, but should we wish to enter our red-tail to both squirrel and rabbit, we should not start with the easier quarry (rabbit) as if they get used this easier quarry, they will reject anything that requires a bigger effort on their part. Squirrels are some of the hardest quarry to catch and a red-tail capable of hunting them, will be able to hunt most prey, especially if she has been introduced to soar hawking.

Going back to the rabbits and other furred quarry, we will not need to make more than a few slips with bagged quarry to enter our red-tail to a particular prey as by instinct, they will be naturally attracted to such quarry. Even so, we should not rule out the possibility of hunting feathered quarry only because we are hunting with a hawk. Red-tails are one of the biggest predators of birds (including duck, pigeon and pheasant) and if we wish to hunt such quarry and it is available in our area, we should definitely go ahead. In fact, Bandit's first few bagged quarry slips were to pigeons and not furred quarry and he didn't even hesitate for a second and bolted straight after them, crashing through all the bushes and trees in my garden and anything that stood in his way (including me, as they went in between my legs!), till finally just like a goshawk but more savagely, he thrust himself by surprise and fell like a nuclear bomb on the poor unsuspecting pigeon, who fortunately died instantly. I also remember this quite well. Bandit had the typical red-tail (or eagle-like) posture, sitting on the pigeon, with his hackles raised, screaming out loud to me. It was quite an impressive scene! However, if a red-tail is not hungry and is fat, no matter what prey you decide to show him, even at half a metre away (0.6 ft), she will look at it puzzled, or worse still, spread her wings and decide to sunbathe. For their first kill, red-tailed hawks really will have to be on their hunting weight; if they hesitate or take to long to go after the prey, their weight must be lowered a little more. Some birds will initially require to be quite low until they have made a few killings, then gradually they will be able to hunt at a higher weight. Do keep this in mind.

No matter what prey we use, we should try to make these first slips of bagged quarry easy for our red-tail as we need for her to feel confident about what she's doing and not get discouraged. There are many ways to do this, for example, when using pigeons as bagged quarry, falconers sometimes pluck a few feathers from their wings so they cannot fly as fast or as high and also tie their legs together while using a pigeon harness so the pigeon doesn't fly away. Hawks not only need to acquire confidence but also learn where they must grab their prey, though red-tails surprisingly don't need so much coaching in this. For example with hares, they usually do tend to grab them with one foot on the head and the other a little further back on the body, and using all their strength and impact from the attack, these quarry will not be able to put up much of a fight. Also, red-tails are quite large and powerful, and it will be much easier for them than other hawks, like Harris hawks for example to catch certain types of prey and in these situations, technically it will not be too important if they don't grab the prey exactly at the right place. Remember, when entering a bird to any quarry a few baggies will be enough. If we give them too many, they will get used to catching these easy prey and will not want anything else.

If we have decided that we wish to hunt only hare, the most appropriate thing would be to enter our hawk using a hare baggie as rabbits not only look different but also don't behave in the same way. However, it may be difficult to find and catch a hare to use as bagged quarry, so if this isn't possible, we will have to make do with rabbits, but as soon as we can, give her only hare as otherwise she will go for the easy rabbits. *The key when entering a hawk into any prey is to always enter her to the most difficult quarry first.*

5. Basic Training

In Spain, the beginners quarry for hawks is usually rabbit (precisely because it is much easier) and then, they move on to hare, but this will not bring us success in the long run. Falconry is not always easy and if we decide to take shortcuts, such as the rabbits here, we may never get to our final destination. In other words, if you eventually want to hunt hares with your hawk, don't even look at the bunnies.

Birds of prey in the wild do not usually feed or hunt every day and they cannot afford to waste a great amount of energy in getting their prey, so will always go for the weak or easier prey, although occasionally may pursue something which seems to fall out of their reach. If when we enter our red-tail into quarry, she seems more interested in mice for example than rabbit, we can avoid this turning into a real drama by rapidly introducing her to the desired quarry and rewarding her generously each time she catches a rabbit. Should she still manage to catch a mouse anytime in between, do not give it to her or allow her to eat it as they will lose interest in everything else. We have to avoid them becoming lazy and just going for easy food. This is one of the biggest mistakes that falconers make with red-tails and which are impossible to correct later on. Needless to say, under no circumstance should we feed our red-tail live mice or rats (which she will take as a baggie) as we will be encouraging her to focus her hunt on mice. Going back to the bunnies, a few bunnies will not cause irreparable damage, but if what we really desire is to catch large hare, we must rapidly introduce her to this quarry and not let her ever catch a rabbit again.

The day we give her the first baggie, she will need to be keen and on hunting weight. We will also need a friend to give us a hand. She must not associate in any way the rabbit with us, as if she does, again we will be creating a bad habit here and she will always depend on us to provide food for her instead of really going out to hunt. When she gets her baggie, we must get there as soon as possible. We may not need to finish it off for her, but we should not let her eat certain parts of animals (see Chapter 1), such as the head, crop, or intestines and get rid of these as she is tearing into her catch. We will need to let her eat all she wants until she is fed up with these baggies, even if it means not being able to fly for a few days until her weight goes down again. Never forget again that red-tails are like gluttons (it may help to have a mental image of Garfield the cartoon in mind here), so this is the way to make them learn and really remember.

Going back to making in and collecting our red-tail from her prey, we will proceed in exactly the same way as we did when picking her up from the lure, and we will always have to give her tidbits. Later on, when we have been hunting for a while and decide to try and catch several prey in a day, we must always no matter what, give her at least a reward for each one and if it has been an exceptional slip, call it a day and let her eat all she wants. In any case, just like with the lure training, we will always finish feeding her on the fist.

The Red-tailed Hawk

Female Buteo jamaicensis with freshly-caught hare in Portugal.
Photo courtesy of Fernando Flores.

6. Hunting with the Red-tailed Hawk

For many falconers that utilize game hawks, prey animals that have fur (mainly rabbits and hares) are pretty much the only species that are thought of in relation to the use of the red-tailed hawk. This is the case because the red-tail continues to be seen mainly as a typical hawking bird, and its capacity to fly at greater heights is generally ignored. This is the case for many species that have been identified as "game hawks" and especially holds true of the red-tailed hawk, which continues to be largely unknown, not only among the wider public but among many falconers in Europe with regards to her potential. It is essential that we liberate ourselves from this way of thinking and that we open our eyes to new possibilities. The red-tailed hawk is a bird that represents the future of falconry—it is the key that will unlock the door to a new and exciting world of great possibilities and hunting adventures. We only need to give it a chance to open this door, and to enjoy everything that we have previously denied ourselves.

As discussed previously, red-tails can be flown using traditional hawking techniques, but they can also be used for soar hawking. Thus, in contrast to other birds used in falconry, these hawks may be used to hunt animals other than mainly just rabbits and hares. This having been said, the large majority of falconers continue to use red-tails to hunt only furry prey, especially hares. For this category of animals, red-tails are clearly the best option. There is nothing wrong with choosing the red-tail for this quarry, as this is really what the red-tail is best suited for, however, I would like to make a number of remarks before you decide to focus on just one sort of prey in order to make things easier for you and before you disregard any possibilities that red-tails may have for hunting feathered quarry.

Hares and rabbits

In Spain, as in the rest of Europe, the most commonly hunted prey by game hawks are hares and rabbits. In contrast, the most commonly hunted prey in North America are jack-rabbits, cotton-tails, and squirrels. All of these species are ideal prey for red-tails, since these hawks have a natural instinct to hunt such animals. They thus do not require a great deal of training in order to learn how to do so and seem naturally attracted to them by instinct.

Hares and jack-rabbits

Although in Spain as in the rest of Europe, the red-tail made its appearance just a few years ago, she is still not yet widely used in falconry as much as in the US—where the red-tail seems to have dethroned the goshawk as the bird of choice for falconers who hunt hares or jacks—though falconers in the United Kingdom have also flown them for the past two decades to a smaller extent at both hares and rabbits.

The Red-tailed Hawk

Although to the untrained eye, hares seem to be similar to rabbits, they are in fact very different to them. It is for precisely this reason that very different hunting strategies need to be employed for these two related but distinct animals. Physically, hares not only have a longer and larger body (with some of the largest specimens being able to weigh up to 5–6.5 kg / 11–14 lb), but they have much longer hind legs and longer ears that have black tips, as well as larger feet.

In Europe, the most widespread hare is the European or brown hare (*Lepus europaeus*) which weighs on average of 3–4 Kg (7– 9 lb) - with some individuals reaching 6.5 Kg. (14 lb)- and can be found throughout northern, central and western Europe, including most of the UK. It breeds on the ground rather than in a burrow and relies on speed to escape, and is the UK's fastest land mammal, capable of reaching speeds of up to 70 km/h (45 mph). The brown hare is replaced in the northwest and western highlands in Scotland by the mountain hare (*Lepus timidus*), possibly introduced by the Romans. The latter is easily distinguished by its white tail and is largely adapted to polar and mountainous habitats like the Scandinavian region. Weights range between 2 to 6 Kg (4– 13 lb).

Spain is actually home to three species of hare: the above mentioned European hare, the Iberian hare (*Lepus granatensis*) and the broom hare (*Lepus castroviejoi*). These are geographically distributed throughout our peninsula, the most common being the Iberian hare. Of these, the largest is the European hare.

Falconers in the US, can find four species of jackrabbits; the antelope (*Lepus alleni*), the white-sided (*Lepus callotis*) and the more common white-tailed and black-tailed jack rabbits (*Lepus californicus*). Of these, the most commonly hunted is the black-tailed jackrabbit and is the third largest North American hare, second only to the antelope jackrabbit and the white-tailed jackrabbit which weighs on average 3–4.5 Kg (7–10 lbs). These jackrabbits can weigh between 1.5 and 4 kg (3–6 lbs), are quite fast, reaching speeds of up to 64–72 km/h (40–45 mph) and can leap 6 m (19 ft) in the air in a single bound.

Going back to hares and their behaviour —unlike rabbits—will fight back against red-tails if they are attacked. They will bite, scratch, and kick. Hares are capable of inflicting serious—even fatal—injuries—on red-tails. I would recommend that " squirrel chaps" be used to protect red-tails against injuries that might occur as a result of hare or squirrel bites (see chapter on "Equipment"), since one of the main means of defence of hares is to bite. Many red-tails have been severely injured by such bites.

Secondly, I would make the general recommendation (with respect to these and to other prey) that red-tails be kept in optimal physical condition: strong, well-muscled, and well-nourished (which is not the same as fat). This is important, because of the popular but ill-founded belief that a bird maintained in a state of semi-starvation will hunt properly. A bird needs all of its energy to hunt this kind of prey.

The pursuit of hares may on occasion take longer than expected—or longer than we would like. However, fit red-tails are able to pursue their prey for very long distances without any problem. The success of a pursuit on any given occasion will depend on the stamina and the physical condition of the individual red-tail specimen. One way of assuring that your red-tail is in proper condition is to fly her every day, to regularly induce her to jump vertically, to train her to fly from the foot of a hill upward, and to fly upwind.

Finally, although red-tails are more than capable of handling such prey, we must do everything we can on our part to help our hawk at the time she captures a hare, especially in the case of younger and more inexperienced birds (i.e., birds that have not yet learned how to properly grasp a hare; by its head). For this reason, bagged quarry are usually very useful, as they can teach a hawk where to aim for, although they are not permitted in every country. In any case, as soon as our hawk makes a capture, we must get to her as soon as we can in case she needs assistance.

As is the case with the majority of the red-tail's prey, hares or jacks can be hunted from either greater or lower heights. However, it will usually not work to have the red-tail fly directly from your fist—unless you are already atop a hill. If the bird is not flying from high above ground level, it will be indispensable to give the bird some kind of height advantage to the bird, either by letting her place herself on a perch or some similar place while prey are flushed, or using a T-shaped perch (which can also be used to hunt rabbits in the open field) of the type commonly employed in the United States and Mexico.

Both hares and rabbits like to run. In contrast to rabbits, hares are extremely fast and very difficult to catch; they also don't usually look for hiding places when they are being pursued. Instead, they usually run with the wind in order to throw the red-tail off their trail. They will also sometimes try to run uphill, since this is something that usually makes things more difficult for the bird that is in pursuit. For this reason, a fit red-tail is a must if we are to pursue this quarry and we must also be able to launch her in the direction of the wind. In this way, she can build up speed for the attack. This together with the height, will be our best advantage.

Rabbits

As far as I have been able to see, in Spain, the United Kingdom, and in the rest of Europe, the rabbit is usually the prey of novice hawkers and are ideal for red-tails. Besides not always being easy to find by inexperienced falconers, this quarry can also present certain problems for falconers who use rabbits as an introduction to hunting when their final objective is really the hunting of hares. All birds of prey have a "lazy side" in that they adapt their behaviour to prey they are used to. This then makes it difficult when they are faced with more challenging prey. ***Instead, you should always introduce your red-tail at an early stage to the most challenging prey.*** Once the red-tail learns to hunt more difficult prey, she can easily learn to hunt easier prey. However, the reverse does not hold true.

Though similar in appearance, rabbits can be clearly distinguished from hares in that their young are born blind and hairless. Hares, instead, are generally born with hair and can see from birth. Another big difference is that all rabbits (save for the American cottontail rabbit) live underground in burrows or warrens, while hares live usually rest on grass areas above the ground (as does the American cottontail rabbit), and usually do not live in groups. A third and physical difference, is that though similar, hares are generally larger than rabbits (almost twice the size!), with longer ears, and have black markings on their fur.

However, like hares, rabbits are also mammals that belong to the Leporidae family (but to the Lagomorpha order) and can also be found geographically distributed throughout the world. The best known species in Europe is the European Rabbit (*Oryctolagus cuniculus*), native to southern Europe and weighs approximately 1.3–2.2 kg (3–5 lb). This is also the most social rabbit, sometimes forming groups in warrens of up to 20 individuals and the only rabbit to be domesticated. In Spain, we have the Iberian rabbit (*Oryctolagus cuniculus algirus*) which can be found and has an average size of 600g to 1 Kg (1–2 lb).

In North America, there are many different species of rabbit though the most commonly distribute and hunted in falconry are cotton-tail rabbits (Sylvilagus genus, of which there are 16 lagomorph species). Cotton-tails are very similar to the wild European rabbit described above and their weight ranges from 800 to 1,533g (1.8–3.4 lb). During the day, these rabbits often remain hidden in vegetation. When spotted or under threat, they run in a zigzag pattern, sometimes reaching speeds of up to 29 Km per hour (18 mph). Though rabbits do not usually put up a fight like hares do, they will struggle and do their best to get free. This may result in a small tumble with some broadwings though generally, it should not cause too many problems (if any) with a red-tail. Again, just like with hares, our hawk should learn to grip this quarry properly and tightly, though of course, if not perfect the first few times, this will improve with experience.

In general, most rabbits can be found in prairies, in areas near cultivated fields, in abandoned areas, and in areas that were previously farmed (rabbits are not found in areas where there is no vegetation at all, as they prefer grassy, weedy areas). Jut before a big storm, they can often be seen feeding (so that they will not be in a position of having to search for food when they are unable to obtain it) and are therefore also quite vulnerable after a spell of bad or cold weather. When first entering your hawk to this quarry, make sure to take her to an area with abundant rabbits so they can easily be flushed and she doesn't get discouraged. Bunnies are not too difficult a prey although finding them is sometimes tricky and if they have been hunted down in certain areas will possibly outsmart our hawk. The best hours to find them are around dusk or dark but should this prove not too fruitful, we can also try lamping (where legal,

6. Hunting with the Red-tailed Hawk

please consult with local authorities and clubs) as rabbits are mostly nocturnal and it works quite well with broadwings. However, I wouldn't recommend that beginners attempt to enter their hawk on their first rabbits through the use of lamping. This is best done when the hawk has a little experience and is used to hunting as we could also risk her getting lost and in the dark, it will be much harder for us to find her.

There are many different opinions regarding the proper method of hunting bunnies with a red-tail. Some believe that they should always be flown from the fist, while others believe that they should be flown freely (soar hawking or following on). I think that whatever the technique in mind, we will first need to examine the terrain and available prey and keep in mind the fact that each situation is different; it will not always work to employ the same strategy. Like many falconers, I am also of the belief that, generally speaking, a well-trained and properly conditioned red-tail will always be more effective flying freely and being placed at a particular vantage point, where she can decide when to pursue a given quarry and how to pursue it. She will always have a greater advantage when she can fly from a greater height and stoop down: it is usually not possible for a red-tail to do this when she flies from the falconer's fist, although there are, of course, exceptions to this rule.

Wind is a very important factor, as Ben Long correctly pointed out in his article in Chapter 3. In fact, wind may be the key to success to flying the red tail, and may serve as the bird's greatest ally. Whenever possible, you should try to fly your bird against the wind or upwind to give her a good work out and help increase her agility. Going back to the bunnies, they will usually try to go upwind and if your hawk is in top condition, this may not be a problem. However, if you can help your hawk by finding the most appropriate hunting perches upwind, you will be assuring a great and easier slip; forcing the bunnies to go down wind and giving your hawk the added advantage of extra speed.

Basically, when hunting rabbits you should whenever possible to set up a slip where like with ducks (see feathered quarry), the rabbit can only run away from the red-tail in an angled direction. As with most quarries, it will be more difficult for the red-tail to capture game at a close distance though not impossible). Strange as it may seem, it is actually easier for red-tails to capture rabbits when they are running away from them. Whenever possible, as I mentioned above, it is always better to arrange things so that your red-tail flies downhill. Not only is this physically easier for the bird, but this affords it a better line of vision.

Also keep in mind that most rabbits will not be willing to cooperate for obvious reasons and will usually try to run against the wind. Last but not least, be careful when chasing rabbits with fences as they tend to dive under them and the red-tail may not even be aware that it cannot ("yet") go through a fence!

In areas where there are few trees like in the south of Spain or open plain areas, you can still use the only tree available and try to force the rabbits in that direction. You can also use a T-perch made of wood as a resting place for your red tail like the ones used in the USA and Mexico. In order for this to work, you have to train your red-tail to come to the perch by using food as an inducement: she will get used to it in no time and is a tactic that usually works well. Needless to say, you can also use soar hawking where there are no trees for hunting bunnies.

Note

When flying your red-tail freely, and not from the fist, many falconers complain that they don't feel part of the team, or worse still that they feel they are losing touch with their hawk. To avoid this, we should also make sure that the red-tail does not lose the habit of returning to your fist when you let her fly freely from trees. For this you should call her to your fist at least twice every time you take her on a hunt. If she does not comply, you need to use larger pieces of food as an inducement. I also think that it is critically important that, after the end of a day of hunting or training, you end it by giving your red-tail something to eat on your gloved hand (calling her a few times, alternating food with no food, so that your hand comes to function as a conditioned stimulus, enabling you to summon your bird even without food. You must always end the session by letting her feed on the glove till she has a full crop.

Squirrel hawking

In recent years, squirrel hunting has become very popular among falconers in the United States, and the hunting of squirrels poses one of the greatest challenges for hunting birds—red-tails included. If red-tails are to be used to hunt squirrels, their claws should be very sharp, since the skin of squirrels is very hard and they will need sharp claws in order to be able to seize them properly. It is indispensable for the falconer to be aware of what is going on, and to be ready to help his bird when needed (but never by using a knife to kill the squirrel, but rather by the "squeeze". It is not necessary to add that the red-tail should fly without its field jesses as she could get tangled up on a branch. She should also be fitted with chaps for added protection (see Patterns for these items in the appendix at the end of this book) although it is possible that she may still be bitten while wearing these. Finally, the red-tail should be in optimal physical condition for hunting squirrels, since this is a very demanding activity.

In Europe, the most common European squirrel is the red squirrel or Eurasian red squirrel (*Sciurus vulgaris*). With an average size of 19 to 23 cm (7.5–91 in), and a tail length of 15 to 20 cm (5.9–7.9 cm), its weight ranges from 250 to 340 g (0.55–0.75 lb). Though far from being considered as being an endangered species world-wide, in recent years the populations of red squirrel have declined. This being particularly true in the United Kingdom and in parts of central Europe where they are under threat due to the introduction of the grey squirrel (*Sciurus carolinensis*) in the late 1870s from North America. Do note that it is actually illegal to intentionally injure, catch or kill them, disturbing their dreys or cutting down trees with dreys in them, not to mention any other mode of pursuit. The red squirrel is now protected in most of Europe, as it is listed in Appendix III of the Bern Convention; it is also listed as Near Threatened on the IUCN Red List. Additionally, they have been protected by the law in the UK under Schedules 5 and 6 of the Wildlife and Countryside Act.

The eastern Grey squirrel (*Sciurus carolinensis*) is slightly larger than the red squirrel with a length of 25 to 30 cm (9.8–11.8 in) and weighing between 400 and 800 g (0.88–1.8 lb). Grey squirrels are widespread and not protected in the UK. Unlike the red squirrel, due to their large number they are considered a pest of forestry or plague, and are often killed on roads.

Back home, in Spain, we can find both the red squirrel and the Barbary ground squirrel (*Atlantoxerus getulus*), which is limited to the island of Fuerteventura, where it was introduced in 1965 from Africa. This is a smaller squirrel with a body length 16–22 cm (6.3–8.7 in), tail length 18–23 cm (7.1–9.1 in) and average weight range of approximately 300–350 grams (0.66–0.77 lb).

The three most common species of tree squirrels in North America are the red (*Sciurus vulgaris*), grey (*Sciurus carolinensis*) and fox squirrel (*Sciurus niger*). The latter is sometimes also known as Raccoon Squirrel or Monkey-faced squirrel and is the largest found in North America (nearly double the size of the grey squirrel). The fox squirrel measures 45 to 70 cm (17.71 - 27.55 in), with a tail length of 20 to 33 cm (7.87 – 13 in), and can weight anything from 500 to 1,000 grams (1.1 lbs - 2.2 lbs). These squirrels spend more of their time on the ground than most other tree squirrels, though they are nevertheless agile climbers. It is only legal to hunt the grey squirrel and fox squirrel in the US.

For most hawks, squirrels can be one of the most difficult quarries to ever pursue and some may only catch it occasionally. Red-tails, however if properly trained, fit and experienced are very capable of this quarry and can provide very entertaining (and successful!) hunting outings with plenty of game in the bag.

Squirrel hawking sounds like a lot of fun but there are also disadvantages to it. Both falconer and hawk must be extremely fit and agile, as you may end up climbing many trees. Squirrels are small enough for your hawk to eat wherever she has caught them; this means that you will obviously still have to make in and avoid your hawk from feeding up and flying away afterwards. You will definitely make use of your tree climbing skills here, so if these are a little rusty or getting high up in a tree could be a problem for you, I would reconsider this type of hunt.

Squirrels are also very aggressive and your hawk will need extra protection (see equipment section for squirrel chaps). If we are not careful, what could be an

The Red-tailed Hawk

afternoon outing of fun could turn into a disaster where your hawk is crippled for life or where you, also may receive a nasty injury. Squirrels will not hesitate to bite any of you and when your hawk catches one, you may need to run in quickly to help her finish the squirrel off safely.

When hunting squirrel, generally red-tails will try to make the squirrels climb up the tree branches as high as possible, since the upper portions of trees afford little protection (and squirrels are less able to move quickly on shorter branches than on longer ones). The falconer must then aim to try to keep above his hawk directly above or ahead of the squirrel. The chase will usually begin at the highest point of the tree. If the red-tail cannot get to the squirrel, they may try to force them to the ground, either by laddering down the tree or by pouncing against the squirrel in order to get them out of the tree and onto the ground.

The falconer can help his hawk here by banging the tree bark with a stick and also shouting at the squirrel. Some hawks though, especially red-tails will have no problem in employing other strategies which surprisingly can consist in them literally "smashing" on to the squirrel and often red-tails will swoop at the squirrel in an attempt to grab her (and the tree with it!) as they pass by. Other times, red-tails may come across a squirrel "nest" or drey (commonly known as a meatball in the US by falconers) and we must encourage and shout at her to go in.

Some squirrels, particularly the larger ones, may even confront the red-tail and try to defend themselves. If it is an inexperienced hawk, this could be quite intimidating for her. Other squirrels will freeze in a panic. If this happens, we have two choices. The first would be to be patient and let our hawk decide the best way to get to the squirrel; including making the squirrel even more nervous or trying to confuse it by suddenly changing place or jumping on the ground. The second choice is for the falconer to help the hawk get the squirrel moving, either by beating on it with a stick and shouting, or even throwing small pebbles at the squirrel.

Jim Gwiazdzinski is a master falconer in the United States who has avidly hunted squirrels for many years. It was not until the second half of his second year as an

Red squirrel in the snow. Photo by Geoff Dennis.

6. Hunting with the Red-tailed Hawk

*"Bird", Jim Gwiazdzinski's red-tailed hawk in Rhode Island with squirrel.
Photo by Geoff Dennis.*

Photo shows a squirrel nest or "meatball". Photo by Geoff Dennis.

The Red-tailed Hawk

This squirrel seems to need a parachute! This is what can sometimes happen when the squirrel (mainly the grey squirrel) sees no other way out and decides to "bail out" or jump to another tree or branch and at times even the ground. Photo below shows a squirrel nest or "meatball" Photo by Geoff Dennis.

Here the squirrel doesn't know what's coming! A red-tailed hawk that has no need for a parachute! Photo by Geoff Dennis.

apprentice that he was introduced to "squirrelling", as it is popularly known in the United States. At that time, the practice was not especially popular, because of the risk of squirrel bites. Jim met Steven Aldin, a falconer from Massachusetts, with whom he hunted grey squirrels. After that very first hunt, he was hooked.

Jim spent his first year hunting rabbits, and was unable to see 95% of them (he could see the bird approach the rabbit from a tree, make a modest stoop, and stamp down on the weeds, and could see and hear signs that the bird was pursuing a rabbit. Jim felt that he was losing something, since he did not have the satisfaction of viewing the pursuit and capture of the prey. Thus, for him, squirrel hunting was a more personal activity, since he was able to more closely follow the entire pursuit of the prey by the bird, and it was something the he could also actively participate in.

For those of you who have never tried it, the following article by Jim can give you a little taste of what squirrel hawking is like:

Red-tailed hawks:

The bread and butter of American Falconry by Jim Gwiazdzinski

Deeply woven within the fabric of American falconry is the red-tailed hawk – (Buteo jamaicensis). No other species of bird has graced the glove of the American falconer during his initial apprentice years. For many falconers the red-tail or "RT" continues to be a hunting companion long after the post-apprentice years. While many falconers move on to longwings, shortwings, Harris's Hawks, etc., it is the RT which has allowed the aspiring falconer to delve into the sport of falconry.

Through the years while practicing falconry, there has been one constant in my life - the red-tailed hawk. Because of this bird, my falconry has become more proficient, more fulfilling, and most importantly - a lot more fun. It is the RT and squirrel hawking that has become my bread and butter of falconry. I strongly believe the RT is underestimated, under-utilized, and does not receive the respect it deserves because it has been labelled as a "beginners" bird or worse, a buzzard. With this in mind, one thing has become very apparent; red-tailed hawks only get better as the falconer gets better. The potential of a RT is directly related to the potential of the falconer. These birds have an incredible amount of tolerance, talent and potential that is many times over looked or not exploited due to the inexperienced apprentice falconer. When it's time to go out, catch quarry, and have some heart thumping fun, I can only think of one bird: the red-tailed hawk. As I write these words, I have my four times inter-mewed tiercel on a bow perch in the same room with a foot pulled up. The hawking season just started and we've already had a number of spectacular hunts and a good many squirrels in the bag.

A few comments and observations

Before I ramble on about my favourites way to spend a day – squirrel hawking. I wanted to comment on some observations I have made. I am not going to comment on the manning, training, and free flying stages of falconry. There is a superfluous amount of information written on those facets of falconry. But, I will say this; in order to successfully fly a RT on squirrel (or any falconry bird on any game for that matter), the bird's hunting weight must be determined. If the bird has been properly trained and manned and is at "hunting" weight, game will be put in the bag. Flying weight, or more accurately, "following" weight, is not hunting weight. This weight is not going to put game in the bag. Find the hunting weight and both hawk and falconer will have a successful hunting season. If you enter a woodlot and your bird repeatedly "blinks" (refuses) slips after squirrels, call the bird down. Do your homework and find the bird's hunting weight. You will be wasting your time if you try to force a hunting situation with an uncooperative bird. This too is where bad habits can begin to become the norm and it is the bird who is training the falconer, rather than the other way around. A RT at hunting weight is an effective and intensive gamehawk and has only one thing on its mind - to hunt and kill game. If this is not the case, you need to do some troubleshooting.

During the course of a hunting season, I start my bird at a lower weight during the beginning of the hunting season, and then as the season progresses into colder months, I raise the bird's weight. Basically, I try to hunt my bird as high a weight as I can while maintaining solid field control and a focus on game. What must be kept in mind, is as the falconer gets deeper into the hunting season (January and February for American falconers), a RT will gain a great deal of muscle weight. Squirrel hawking by nature is an extreme workout for the hawk. If the falconer climbs trees, well then, he too gets a good workout as well.

A common mistake I have seen, time and time again, are falconers not factoring muscle mass into the equation. The falconer weighs his bird, finds the bird above the usual hunting weight and immediately thinks the bird is overweight and needs to be lowered to hunting weight. This should not be the knee-jerk reaction of the falconer. A hunting hawk is not a machine. A falconer cannot simply dial a hunting weight in and expect the same hunting performance each and every time. A falconer must adjust the hunting weight according to the performance of the bird. The result will be a strong, fit, successful hawk. Keep in mind weather, quality of food, and how often the falconer hunts his hawk is directly attributed to the performance of a hunting hawk. For instance, a hawk given pigeon breast, as opposed to rabbit meat, will hold the weight a lot longer due to the high quality of the pigeon meat. Factor this in with a warm front, and guess what? You've got yourself an overweight bird that will most likely sit in a tree and watch game run by. The keenness of the hunt is gone, the bird is not "sharp" enough.

A risk to any squirrel hawk is the chance of the gray squirrel (Sciurus carolinensis) delivering a nasty bite. Because of this very fact, I administer the squeeze. When the bird has come down out of a tree with a squirrel I run in, grab the squirrel, and squeeze the chest cavity. The squirrel can no longer bite the hawk and is more or less suffocated at the same time. I have control of the final and most risky hunting moments. From day one, I make in on the bird and assist the bird with the squirrel. I have yet to have any adversarial behaviours relating to this approach.

Hunting squirrel with the Red-tailed Hawk: My own bread and butter

Squirrel hawking is a rich alchemy of visually thrilling flights. Because a variety of flights are always at hand, and the "RT" must put the cagey squirrel at a disadvantage, the bird is required to "think" its way through a hunt. Squirrel hawking birds are incredibly fit due to the vertical flights that must take place in order for the hawk to get above the squirrel. A close hunting bond is established (especially when you climb trees for your bird) and, well – it's downright fun. Squirrel hawking can be a ground chase, it can also be a RT crashing into a leaf nest (a meatball). A corkscrew flight can take place with the squirrel running down the tree as the RT corkscrews after the squirrel. Those flights are fast and precise, with the bird avoiding branches and crotches while in a stoop. There are mid-air snatches, which happen to be one of my favourites. The RT literally grabbing the squirrel out of thin air as it tries to jump from one tree limb to another.

Aside from the aesthetics of squirrel hawking, pursuing "ole' bushytail" is a very straightforward undertaking. Remember, this is with the assumption that hunting weight has been determined. After the bird has been manned down, trained, flown on the creance, and free flown, the best part of the falconry process presents itself: hunting and/or squirrel hawking. In order to get a newly trapped and trained RT versed in squirrel hawking (if it hasn't been hunting squirrels in the wild prior to trapping) with a falconer, go to one of your "honeyholes," (or best hunting spots) put the bird up in a tree, and walk the woodlot. Or, if you visually spy a squirrel while driving in your hawking area, stop the vehicle, get the bird out of the giant hood or hood, put him up in a tree and start pushing the squirrel. Simple stuff.

The visual acuity of squirrel hawking is an added advantage for both the falconer and the hawk. This is also why squirrel hawking is so much fun. Making sure the hawk will have flights on squirrel is paramount. The sooner a squirrel kill has been made, the sooner the falconer is going to reap the benefits of a seasoned bird. Success breeds success.

Here in Rhode Island, we have the gray squirrel. The gray may be common, but this does not mean squirrel hawking is easy, nor a squirrel will always be put in the hawking bag. Squirrels are smart and extremely athletic. They also bite, roll in a ball like a cat and kick, can run straight up and down a tree, jump from one limb to another, hide in brush like a rabbit, take cover in a tree hole or a leaf nest, and bail from a tree from more than 40 to 50 feet (12–15m), hit the ground and run into cover. Squirrels are tough. When a squirrel is on the move, it is on the move for a reason. A squirrel always has a plan. Keep this in mind while running through a woodlot with the squirrel running the limbs overhead. Know where the escape hatches are and try and head off the squirrel.

During my apprentice years, I beat a lot of brush for rabbit slips. My former sponsor primarily rabbit hawked, so you guessed it, I too did a lot of rabbit hawking. I enjoyed the hunts, but it was devoid of one essential factor; witnessing the chase and capture of prey. Many rabbits that are caught by a RT here in New England are caught out of sight of the falconer. Usually the falconer "hears" the rabbit caught, without ever seeing the capture of prey. Briars and thick cover can keep the falconer out of the loop. If you're running dogs, you can be even further from the action. Squirrel hawking is completely at the other end of the spectrum. Most squirrels put in the bag by both the hawk and the falconer are visually witnessed by the falconer and the hawking party from beginning to end. This is why I cannot get enough of squirrel hawking. This is also why it is both exciting and fun.

Now, I'm going to confess something here. I climb a lot of trees for my bird(s). What I have found from my tree climbing antics is an extremely strong bond between myself and the RT's I have flown. The bottom line is the birds associate me with a slip and usually a squirrel in the bag. This of course does not mean that the only way to successfully fly a RT on squirrel is to climb trees. Sometimes by slapping the side of the tree, the squirrel moves enough for the hawk to jump start the chase. Walking to the side of the tree the squirrel is on can push the squirrel to the opposite side of the tree and where the bird can again get a visual fix. This is when the bird should engage and the chase can begin.

A Few Choice Flights to Leave You With

Allow me to take you away from falconry for a moment. Before you follow me into the woodlots of Rhode Island, USA, let me describe my surroundings. Where I live, I am no more than five minutes from the great Atlantic Ocean. In this body of water we have shark, tuna, and gamefish such as striped bass and bluefish. I spend a great deal of time on the ocean chasing bass and blues as well as a tuna or two. We have gorgeous

6. Hunting with the Red-tailed Hawk

Jim making in on Bird's squirrel. Photo courtesy of Jim Gwiazdzinski.

Red-tail on squirrel. Photo courtesy of Jim Gwiazdzinski.

The Red-tailed Hawk

*Double the fun! Jim and his red-tail bag a brace of squirrels.
Photo by Dave Martin.*

summer days, accompanied by a sea breeze, making a hot, humid, summer day, more tolerable. Waves, originating from across the ocean, possibly from the likes of Portugal, finally meet land and crash into our beaches. What I also have are excellent and fecund hawking grounds for the RT's I have flown.

American falconry is usually portrayed in magazines and books, with grand pictures of big sky and big country: The American West; longwings stooping at extreme speeds on waterfowl and grouse. Tail chases continuing beyond the eyes reach. A sight to behold indeed. But, falconry is also practiced within the confines of woodlots with tall oak and maple stands. Tucked away within those woodlots are falconers chasing squirrels with RT's. Here are a few of those days.

We had been pushing a squirrel for the better part of 30 minutes. There were repeated attempts by the bird to rake the squirrel off the side of a large oak tree, but to no avail. The RT tried to make flights at the squirrel that quickly dashed to the other side of the tree. At one point, for reasons only the squirrel could understand, it changed tactics, running up the oak tree, jumped to a tall Pitch Pine, and buried itself into a leaf nest at the crown of the tree. Four seasons of squirrel hawking taught my tiercel well. The RT launched from the oak tree, flew straight into the leaf nest, punched its feet forward into the centre of the nest pulling the squirrel out. Both squirrel and hawk came parachuting to the ground. I made in, squeezed the squirrel and a squirrel was put in the bag. I have also watched my RT land on top of a leaf nest and continue to pull leaves until it unnerves the squirrel. Sometimes the squirrel is caught right then and there, and then there are times the chase begins yet again and we are off chasing both squirrel and hawk.

One of the more common flights, but certainly one I have not grown tired of is the laddering up by the RT. This is where the falconer can observe the RT "think" its way through the hunt. I'll continue to slap the side of the tree to keep the squirrel within the tree if I don't want a ground chase, (A ground chase is one of the more risky pursuits of squirrel. The hawk will grab the squirrel by the hinds, allowing the squirrel to fold up back around and possibly get a good bite on a hawks toe) while the hawk ladders his way up and above the squirrel. The bird is constantly looking up towards the squirrel. Sometimes my bird will fly to an adjacent tree to gain both height and advantageous angle. Not until the squirrel is put at a disadvantage does the hawk commit to making a flight at the squirrel. If the bird is able to grab the squirrel on the first attempt then obviously the hunt is over, but if the squirrel outmanoeuvres the hawk, the hunt continues.

Myself, some fellow falconers, and hawk walkers were hunting one of my honey holes. The hunting ground has a prolific amount of squirrels. As we made our way through the woodlot, we marked a squirrel that was on the move. We yelled the game call. The bird zeroed in on our call, flew in tighter to the hawking gang as the squirrel ran the tree tops and into a massive oak tree. Locking in on the squirrel, the bird flew to the top of the oak, immediately pushing the squirrel down the tree, corkscrewing with the squirrel more than half the height of the tree. The hawk missed and the squirrel circling around and headed back up to the top of the tree. The hawk continued through his pass at its quarry and flew to another tree. Immediately upon landing on a limb, the hawk wheeled around and flew back to the oak tree. The bird landed a bit below the squirrel and watched as it made its way onto the thin branches at the very top of the tree. The squirrel froze upon reaching the top branches and would not move despite our best efforts. The bird too seemed to freeze. Contemplating whether I should climb the tree, the bird made it an easy decision when he flew from the tree, made a quick half circle and flew straight at the squirrel. Wham! There, more than 100 feet (30m) above us, the RT had the squirrel. Down floated the hawk with a squirrel in his grasp. Exciting stuff.

There is a particular flight, regardless of success or failure, which is hard to surpass. I speak of none other than the mid-air grab by the hawk as the squirrel is trying to jump from one branch to another. There is a sense that time stands still as the squirrel jumps and is momentarily at the mercy of a hard-charging hawk. Standing underneath the chase and wondering if the hawk is going to get the squirrel at that very moment is suspenseful.

The squirrel had been running through a Pitch pine stand, which is tough hunting for the hawk. The pines have clusters of thick, stiff pine needles that can keep a hunting hawk from getting a foot on a squirrel. Squirrels know the game and how to be evasive running the tops of these burly pine trees.

As the squirrel was running and jumping from pine to pine, it made one fatal mistake. It tried to jump onto the reaching branch of a maple tree. Following the flight from underneath, we all seemed to pause as the squirrel jumped. Just as the squirrel came within inches of grabbing the branch, the RT came from behind and snatched the squirrel from mid-air. When the bird came down with the squirrel I ran in, did the squeeze, and gave the bird his rightful reward, warm squirrel meat. We stood there with smiles on our faces as the bird broke away meat, walking out of the woodlots and to our cars and trucks, we couldn't help but relive the flight. It was that good.

In order to truly get a sense of the excitement of squirrel hawking it must be witnessed. I can try and relate my experiences of squirrel hawking, however, as is the case with many things in life, and falconry specifically, it must be lived and experienced first hand. Falconry is a hand on endeavour. I have received e-mails and comments from many falconers who have seen squirrel hawking and cannot get over how much fun they had watching a squirrel and hawk try and outsmart each other. I have been told many times; "*If I knew how much fun it was, I would have started squirrel hawking a long time ago."*

There are more than enough squirrels to go around. With development and hawking covers loss to this development, quarry that is habitat-sensitive, such as rabbit and upland game, is also lost. Due to this encroachment, I believe squirrel hawking will gain importance and its rightful place within the sport of falconry. Squirrels adapt very readily and successfully to the impact we have on our environment. It is because of the red-tailed hawk and the prey we hunt, the gray squirrel, that my falconry is fulfilling and fun. Within squirrel hawking, tucked away are those hard-earned learning experiences, challenges, obstacles, and many laughs. I hope your falconry is as fulfilling as mine has been. From the woodlots of Rhode Island, with a squirrel or two in the bag – I wish you success and a hawking bag full of game.

Some Lasting Thoughts

It's been a few years since the Spanish version of this book came out. I want to thank Beatriz for having both myself and my wife at the book opening in Madrid Spain. The experience was phenomenal. Meeting falconers from Spain, England, and other reaches of the world will stay with me for a lifetime. Thanks again Beatriz.

With that being said, I feel I must comment on the following; the popularity of squirrel hawking through the years and "pet keeping". A little over a decade ago, squirrel hawkers were hard to find. However, as the years go by, as they always do, I have seen an incredible surge in squirrel hawking. Year after year, there seems to be more falconers pursuing squirrel. My old-time sponsor who has flown goshawks on rabbit and duck for decades is now also flying a RT on squirrel. Rabbit cover is constantly being gobbled up from development every year, making squirrel hawking more realistic and practical if a falconer wants to put game in the bag. This brings me to my last thought – "pet-keeping". What I think cannot get lost is the "game in the bag" mentality. If a falconer is not out hunting and killing with his/her trained raptor, the falconer is pet-keeping. It is paramount that the sport of falconry maintain its capture of quarry mentality. A watering-down effect or a mind-set that not killing with a trained raptor is acceptable - is dangerous. Falconry birds should be killing game, plain and simple. If a falconer finds himself with a bird in the mews more than out in a hunting field or woodlot during the hunting season, the falconer should re-consider the sport of falconry. Falconry needs to be about hunting in order for the integrity of the sport to maintain its identity and not waiver. The sport can easily make a into a watered-down version of quasi pet-keeping/falconry. Falconry must be about hunting. It is not about shows, swinging a lure for an audience, impressing a neighbour with a bird on the fist or impressive falconry nomenclature, nor is it a passing fancy. Again, falconry is about hunting. If the intention of the falconer is to hunt and kill game, the quality of the sport will stay intact.

Feathered quarry

Although the ideal prey for the typical red-tail are rabbits, squirrels, and hares, this doesn't mean that it is incapable of hunting avian prey such as doves and pigeon, pheasants, ducks. Depending on the level of stamina and experience of both falconer and hawk, in time, they may even pursue other feathered quarry such as geese, quail or partridge with a degree of success. In fact, even though the vast majority of the red-tail's prey are mammals, and they tend to instinctively prefer such prey, many falconers use them to hunt avian prey, since they seem well-suited to such a purpose. Red-tails have the ability to adapt and to take advantage of opportunities presented in a hunt.

Pigeon and doves

For many broadwings, these are quite difficult prey and only rarely caught in an opportunistic manner but both Harris hawks and red-tails are perfectly capable of hunting these (both hawks are used by falconers performing pest control services on these birds). In the wild, red-tailed hawks do occasionally prey on pigeon, many performing spectacular flights such as the memorable chases seen on TV with Pale Male, in Central Park (see Chapter 8).

My male red-tail Bandit was briefly introduced pigeons in Spain for pest control purposes through a few baggies and did quite well, though he didn't seem to be as enthusiastic with them as with ducks or furred quarry. If we decide to give this a try, the best way to do so is with a couple of baggies (where legal) or by finding over-populated areas such as those that require pest-control. As always, height will be an advantage for the red-tail but Bandit also managed to catch them bolting directly out of my fist after what would seem an endless chase, where the pigeon would look to find cover and my red-tail would crash right through any obstacle that got between them. If your hawk's fitness levels are up to scratch, the pigeon will eventually tire or be outsmarted and it will definitely be an unforgettable slip.

To catch this quarry with any success, our hawk, must be incredibly fit and if caught, particularly near urban areas, we should never allow her to feed on them as they carry many diseases. Instead when making in, you may replace it with a woodland pigeon (previously killed) whose origins you trust.

Pigeons and doves in general are very fast and will outrun most hawks in horizontal flight, resulting in exciting slips (and a good workout for us!) but if your hawk feels in top shape and is on target weight, she will give it all it takes to pursue this prey till the end. With Bandit, it seemed almost as if it was personal and he wouldn't forget about it until it was over for the pigeon.

Ducks

Ducks can also be excellent quarry for our red-tail though initially not recommended for beginners, as these slips are difficult and require an experienced hawk. With ducks, they are easily scared off or worse still tend to "duck" and dive in the water (with red-tails going right in after them, so expect to get wet!). For safety reasons, the red-tail needs to learn that it cannot capture a duck that is in the water, since the latter will typically dive into the water at the last moment in order to escape. Hunting ducks is something that is difficult for the red tail, and not only do we need a hawk in top shape, but she also needs to be aggressive and daring, and used to flying from a height. Being larger than average in size also helps as well as good sized feet.

The most commonly seen and hunted wild duck in both Europe and North America is the mallard (*Anas platyrhynchos*) and is the common ancestor of the domestic duck. Mallards are quite larger and heavier than other ducks, with sizes ranging from 56–65 cm (22–26 in) with a wingspan of 81–98 cm (32–39 in) and weighing approximately between 750 to 1,000g (1.7 to 2.2 lb); sizes also depending on geographic location as northern individuals will be larger.

As is the case with rabbits and other prey, the red-tail can be introduced to ducks gradually, initially always in shallow ponds. Of course, bagged quarry (where legal) should be used to facilitate this process, as it is in the case of other prey.

Best chance to catch them is basically to get as close as possible to them and then flush them at an angled distance away from the red-tail. Our hawk will have its best opportunity when they are about to take off, before they have picked up any real height or speed. That is why, again, with this prey, the obvious choice will be to fly our red-tail at a certain height and if our red-tail is used to soar hawking, we will greatly increase the chances of success.

The key point here to remember here is to confine ducks to small bodies of water with good still-hunting perches that can provide a good height for our red-tail nearby. In many places, ducks will found in large lakes or deep rivers, as well as in places with no perching posts nearby and this is not at all suitable for our hawk. In order to assure success, we must have the right setting. The falconer should try to flush the ducks in an angled position away from where his hawk is, so that the hawk can attack from behind. If we don't have many trees, we can still attempt to chase ducks, but as we have seen so far, the golden rule with red-tails is to give them the advantage of launching an attach from a certain height. This could be either the top of a tree, but also the top part of a hill or slope, just overlooking a small brook or pond (as was my case back home with Bandit) and let our red-tail fly out from here. The most important thing to remember is that it will never work to move the ducks directly toward the red tail, since the red-tail does not have the same degree of agility as the goshawk.

When a red-tail does capture a duck, it usually does so when it has used the upper part of a tree as its starting point, pouncing on the duck when it is sitting quietly. It is rare for a red-tail to capture a duck when the latter is in motion, but height does afford it an advantage, so that this difficult feat is sometimes possible. It is necessary for the falconer to release the bird toward the trees, letting it find the best perch for its attack, awaiting the opportunity to strike. This is the technique recommended by Bruce Sandstrom a falconer based in the US who is experienced in hunting ducks with red-tailed hawks. He recommends to launch a red-tail into the trees so that she can plan out her strategy and choose the best perch from which to launch her attach. When the red-tail is correctly positioned, the falconer must try to flush the ducks away from the red-tail, so that she can strike at a certain distance and angle.

If we have introduced our red-tail to soar hawking, this will be of great use for duck hunting and will assure greater success in the field and even more exciting flights.

Pheasant

Another prey which can be considered when both falconer and red-tail gain some experience, and if she is introduced to soar hawking, is pheasant. Pheasants are not native to Europe and are mainly reared here for shooting purposes but can be the ultimate challenge in feathered quarry for a hawk and though difficult, it is not impossible and should not be disregarded.

You will need to have a lot of luck and a red-tail that is in optimal physical condition plus stronger than most and with quick feet, as pheasants will put up a very strong fight. Pheasants are also quite difficult to flush and are very fast, much faster than falconry birds though they are not long-distance runners; this is where the stamina of our hawk can be used as an advantage.

Red-tails will have a difficult time in catching pheasants when they are released from the fist—this is, as we have seen, the case with other prey as well. You should therefore fly your red-tail freely, using the technique of "following on" in areas infested with pheasants (please check legal permissions and also shooting seasons), letting your red-tail follow you or allow her to use the technique of soar hawking, which will bring better results. Once again, height will be your ally; it will enable her to attain maximum speed as well as maximum force.

Quail

Although it is also rather difficult to capture this prey with a red-tail, and it is highly preferable to instead utilize the Cooper or Harris hawk for such a purpose. Red-tails however, can sometimes be used to hunt this quarry but if possible, we should use the smaller red-tail subspecies mentioned in Chapter 2 for this purpose as they are much better suited to it and will have a greater chance of success.

Quail are faster than red-tails, both when flying and in initiating flight. When they are frightened, they tend to gather closely and form groups and then after taking off, fly off in different directions. The best way to capture them is through the use of soar hawking or by the red-tail attacking them when they are in bushes. It is possible for a red-tail to catch an occasional quail in flight but this is difficult and will not always be the case.

6. Hunting with the Red-tailed Hawk

The Red-tailed Hawk

*Close-up of "Red Sonya", female red-tail of Manny Carrasco.
Photo courtesy of Manny Carrasco.*

7. Health and Well-being

A red-tail, like any other bird of prey must be in the best of health, to perform with excellence in the field. Fortunately, red-tailed hawks are extremely tough and resistant birds of prey, generally of good health and able to adapt to a great variety of climates. If a few simple guidelines are followed together with regular exercise and a proper diet, it will be easy to keep these birds of prey in optimum health.

Exercise

Red-tails love food and can be the greatest gluttons! Having said this, they do have a tendency towards putting on weight, or rather, accumulating fat reserves. This must be remembered both when being flown and regarding their diet. Initially, it can be a very difficult and frustrating task to lower their weight to begin the manning process and training.

It is important that they receive regular exercise, if possible every day or as often as possible. A red-tail can hardly be expected to perform well in the field, like any other bird of prey, if it is just a "weekender". Like all creatures, red-tails need regular exercise and a good diet. If it is not possible to fly them everyday, we must at least exercise them a little, for example doing vertical fist jumps. This will ensure they do some exercise, build up muscle and keep the calories down!

Diet and weight management

One of the key elements for the practice of falconry and for the well-being and optimum health of birds of prey is a proper diet, followed by a strict but stable weight management (knowing your bird's flying weight) to achieve "yarak". But what exactly is flying weight? This is the key basic principle of falconry and quite simply what it means is that a bird of prey will only hunt (or work for food) when it is hungry. The more the hunger, the more motivated she will be to hunt. Flying weight, therefore is the weight at which our bird will be motivated enough to hunt without being too weak and in starvation. This weight should always be lower than its initial weight or the weight of the bird upon arrival. To obtain this flying weight, we will need to lower the initial or top weight of our red-tail, though not too much (as we should make all efforts possible to always fly any bird of prey at their highest possible weight) to a weight where she responds. The red-tail should be no exception to this rule.

Weight management can be a tricky issue but with a few simple guidelines, it is possible to control it and avoid surprises. Red-tails have a slow metabolism and a tendency to accumulate reserves and therefore put on weight. Normally, the flying weight for birds of prey in falconry should be approximately 10–15% less than its actual fat weight depending also on species and individuals, but with red-tails as with other broadwings, especially in the beginning of their training, it may be necessary to further reduce their weight to 20–25% less. This, however, is quite dangerous and should only be done gradually and for a few days to obtain an initial response and then slowly build up the weight a little. A word of advice: when lowering the weight of any raptor, it is advisable to do it **SLOWLY** and never leave them 1 or 2 weeks without food because we may be in a hurry to fly it or train it. With red-tails, weight management can be particularly frustrating as at times, when even only feeding them just the leg of a 1 day old chick, they may only lose 1 or 2 grams (less than 1 oz). These are the times when we must have patience with the red-tail, but it is always better to take a little longer than to risk the health of any bird of prey. This should also be noted regarding weight management with any bird of prey as health is something that we should not play with. If the bird is flown every other day or after performing exceptionally well in the field and being generously rewarded with a full crop, we must be prepared to spend at least 2 or 3 days without flying our red-tail, just feeding her a leg of a 1 day old chick. An alternative after a full crop from the day before would be to not feed her the next day as her weight will balloon up before we know it.

If however, our red-tail seems to be way below its ideal weight (low condition), we must also not feed her up excessively as she may be too weak to even be able to digest the food properly. Ideally, it is best to give her a larger amount of food distributed throughout the day and use meat that is easy to digest.

The diet is of vital importance for a bird of prey, not only regarding flying weight, but also considering the necessary vitamins and essential nutrients they require. To ensure our birds of prey have the best diet possible, it must be varied (see Chapter 1 and Appendix at the end of the book) and contain bones and feathers, not only for throwing up the castings, but also to naturally shape their beaks, as if these are not used, they could grow excessively and need coping. Even so, it may be necessary to cope their beaks once in a while, as can be seen in the following photograph where Bandit's beak is in need of some coping in order to avoid problems in feeding.

If the beak is not coped when needed, besides making it extremely difficult for the bird to eat, it could cross over. We must carry out periodic checks on the inside of their beaks to remove any leftover food or bones that could produce illnesses.

The seasons of the year should be taken into account as they will affect the diet, weight and behaviour of our raptor. During winter, the red-tail, like other birds of prey, will usually have a lower weight than the weight it would originally have in the wild; if our red-tail has been exercising excessively, it will have lost a great amount of energy at a very fast rate. With sudden weather changes, for example, if suddenly it is colder than usual, our bird of prey will continue to burn more energy only by trying to maintain its body temperature. If our hawk's diet is

7. Health and Well-being

not appropriate, her weight will be too low and she will also not want to fly and could end up falling ill. During the summer, this pattern will be reversed, and birds of prey tend to burn less energy, having even slower metabolisms. Therefore of we are experiencing cold weather but suddenly the weather changes to a hotter temperature, it is likely that our friend may not be in its prime flying weight and may not be hungry. This is the reason why red-tails, like any other bird of prey used for the practice of falconry, should always be flown with a higher weight during winter, decreasing this weight and their calorie intake in summer if they continue to fly during the moulting period.

A last word of caution regarding food. **NEVER** feed dead animals found as road-kill or anywhere (dead by poison) as these may not have died of natural causes and could have been suffering a disease. It is far too easy to assume that animals found on the road have been killed by passing traffic and not because they were not paying attention as they were already ill or feeling off due to perhaps even a poison. Animals that have been hunted with a shot-gun are also not appropriate as food for any bird of prey as they could produce lead-poisoning in our bird. It is impossible to eliminate all shots and many of these can after some time "dissolve" in the carcass resulting in lead poisoning.

Here we can see that "Bandit" needs a little coping. All we need to do is just clip off the excess growth on the tip of the beak and then file/shape (i.e. with a Dremel). Photo by Arturo Gil-Marzán.

The Red-tailed Hawk

The moult

The first moult is usually awaited with great enthusiasm by many of us, particularly when we discover the new plumage of our hunting friend, its new suit for the coming year, which can tell us a great many things on our bird of prey. Red-tails show very similar plumage in both adults and juveniles, the only great difference can be noticed after they have finished their first moult, usually visible during their second autumn, when they are one year old and the red feathers are now beginning to appear on their tail. It is the colour of these feathers which gives the red-tailed hawk its name.

The moult takes place once a year, although not always precisely at the same time every year. It can sometimes appear sooner than planned or be delayed, but generally begins during spring. Red-tails will usually be moulting from the end of April or beginning of May till the end of September, although the length of the moult may vary with certain individuals and can sometimes be a little longer or shorter. A good and rich diet plus supplements can sometimes influence the moult and make it happen a little faster. During this period we must provide the best nourishing diet possible (including rats and mice and vitamin supplements) and must ensure that they are not disturbed or affected by any possible stress as it could affect the quality and growth of the new feathers.

When our red-tail begins to moult, (it will prove very useful to save all the

Bandit's tail during his first moult. Here we can see the new feathers emerging with the typical red hue that gives the red-tail its name. We can also still see the difference with the old feathers (brown with dark narrow barring) of his juvenile plumage. Photo by author.

feathers dropped on the floor that are in perfect state for unexpected "imping" of feathers, should it be ever necessary) it is recommendable to stop flying our bird of prey as in order to moult properly, the weight will need to be raised. The quantity and quality of food will be increased and this together with the heat spells during spring and summer will make the task of flying our bird of prey without losing her impossible, although red-tails seem to not wander off too far whenever they have chosen to do their own thing.

The moult is a time of tranquillity for our bird of prey and they should receive the best diet and care possible to avoid fret marks and to favour optimal feather growth as these will be the new feathers for the upcoming year. However, with many birds of prey, particularly goshawks and red-tails, one can find after the moult that we have to begin manning and training them all over again. Although we must try to bother our hawk the least during the moult, this does not mean that our hawk must be in complete isolation. I highly recommend visiting, talking to and walking with our bird on the fist while going for a nice stroll at least a few times a week even during the moult, so we don't loose our "manning" of our hawk. Some hawks, though fat, will even be quite happy doing a few vertical fist jumps and this will be beneficial both in order to not loose muscle mass and contact with our bird. After I did this with Bandit, he was ready to fly again after the moult in just a few days and he was at a higher flying weight than the year before. In time, red-tails will be able be safely flown at higher weights while being in yarak and performing like no other hawk on the field.

Reproduction

Birds can become extremely uneasy and nervous during the breeding season. Red-tails are known for the aggressive nature and if they are molested or feel too much interference, this can have a negative effect on their breeding. Like with the moult, I recommend a nourishing diet (which will have to be increased and I recommend plenty of mice, rats and rabbit together with vitamin supplements in adequate doses). Also, try and avoid all forms of stress, particularly from outsiders. If red-tails feel threatened they can choose to not breed; breed infertile eggs or worse still, in some occasions, they will kill their young. Make sure if you are breeding that the red-tails have their intimacy. One year, the breeding season was delayed with my red-tails because some kids were bothering them in their breeding chamber. Luckily, they still bred that year but we had to make sure that they were not disturbed.

Although red-tails can lay double clutches, like with most birds of prey, this is not recommended for first-time breeding. Additionally, red-tails should always be parent-reared, therefore, if you are intending to breed them, let nature take its course.

For diet guidelines, nest material to leave in the mews plus additional information on breeding, please refer to Chapter 1.

Health Conditions

Red-tailed hawks, save albino individuals, are very durable birds of prey, however certain illnesses or conditions have been known to affect red-tails and we should do our best to avoid them. Unfortunately, most of these illnesses are caused by improper handling of raptors (many times due to inexperience but mostly due to laziness) and also the fact in itself that hunting is a risky activity. Your hawk can get kicked around in an attack or bitten (hares and squirrels), can also hurt herself while crashing through bushes (thorns, twigs, broken feathers not to mention power lines which can kill a hawk, etc) and can also suffer an illness from eating the actual quarry she hunts.

Excellent hygienic conditions are required at all times and will help reduce risks of your hawk coming down with anything. It goes without saying that your hawk's quarters should be regularly cleaned at least once/twice a week (if possible I would do it every day, even more so if you have more than one hawk). If her lodging area is not clean, she could actually infect her own food through her own mutes and fall ill. Again, this is also why she should have clean fresh drinking (and bathing) water, every day. The hawk quarters should also be thoroughly disinfected at least once a month with an appropriate disinfectant for animals. The equipment you handle every day should also be clean (as well as your hands) and if you have birds which you may suspect are ill, you should use separate equipment for these and keep them in a special quarantine zone. Also, any new bird that you acquire, for safety reasons, should also be kept in a quarantine area until we are sure that health wise they are OK and have also settled down. Disease can sometimes spread quite rapidly amongst birds of prey, and these measures can help control the speed at which it spreads or avoid it. A last obvious thing to keep in mind is that food should be clean and properly thawed as well as prepared on a clean surface. Never keep old food, even if your bird hasn't eaten it all as it will go off very quickly.

Your hawk should be weighed daily and you should note this down in a log book (which can also help us to see her appetite, diet and if there have been dramatic changes weight-wise which could point to health problems). Always keep a look out for any sudden change in your hawk's behaviour or body language as well as her level of activity. Check the insides of beak and foul breath, or cuts in the mouth which could become infected, as well as for a change or loss of voice. When you clean out her quarters, check the colour and appearance of her mutes and castings (is the crop emptying and how often does throw up the casting).

A healthy bird of prey will be alert to everything around her, she will oil her feathers and enjoy being outside and will also seem quite playful. A hawk that seems sleepy an not really with it could actually not be feeling too good, so keep an eye out for that and be aware of any unusual signs which could potentially

7. Health and Well-being

mean an illness or a condition in your hawk. She should also be eating regularly and show interest in food. A lack of appetite or loss of interest in prey could also be a sign that there is something wrong. Last but not least, her eyes should be clean and bright and there should be no discharge from them or the nares and her feathers intact. Should you notice anything different, you should describe such change in behaviour to a veterinarian and consult whenever any doubt or suspicion of illness arises.

I also recommend to regularly look for wounds and to wash your hawk's feet especially after a hunt, checking to see for bumblefoot or cuts which could develop into infections. Also, after a hunt or just once a week, have a look for any broken feathers that may need imping. Beaks and talons should also be checked weekly for cracks or excess growth where coping is needed. If we are not used and unsure about coping, we can consult a veterinary surgeon. If its just a little coping that is needed, we may use a "nail clipper" for cats or dogs with great care to only cope the tip, as if we cut off too much it could produce splinters. This can also be used to clip off the excess growth from the talons. The beak would then have to be filed, or shaped using a power tool (Dremel for example), always hooding our bird and placing it safely within a casting jacket, or wrapping it with a blanket to ensure we do not damage the bird or any of its feathers.

Most of the health problems above can be avoided with a proper diet, good regular exercise and careful regular check-up of our bird. Needless to say, it is of vital importance to ensure that the bird is kept under maximum hygienic conditions and all areas must be regularly disinfected with care.

A last word of advice is to always have your vet's name and number handy and if you don't have one, be sure to immediately look for one in your area, in case you ever have an emergency.

I have included a list of common diseases and ailments below which could affect your hawk so the reader may be aware of their symptoms and as an informative guide. Please do take a few minutes to study it and if in doubt or notice your hawk is "off", contact your vet as soon as possible.

For additional health tips and information including an international list of vets can be found on our new falconry portal for the international community, **www.yarakweb.com.**

Diseases and parasites

Although red-tails are not usually susceptible to disease, except albino individuals, it is a good idea to run through a list of possible illnesses that could affect them and know how to avoid them. At first symptoms and if in doubt, you must consult a licensed veterinarian for diagnosis, treatment and how to proceed.

Avian Trichomoniasis / Frounce

Also known as trichomoniasis and highly contagious, this is a lesion or sore in the mouth and throat, usually seen as a coloured coating on the tongue. A disease of the upper digestive tract usually, it is caused by a protozoan called Trichomonas which is frequently present in the crops of pigeons. This is why when feeding pigeon to our raptors, heads, crops and intestines should be discarded. It is easily and very quickly cured by oral administration of Flagyl, Metronidazole, Entramin, Emtryl, or Carnidazole.

The first signs are flat cheesy plaques inside the mouth. The bird most likely has difficulty while eating food or finds it almost impossible due to the growths impairing the tongue. A loss of appetite could also appear and the mutes could appear to be green.

Aspergillosis

A fungal (Aspergillus fumigatus) infection of the lungs and quite lethal to birds of prey, it can be a result of unhygienic or damp conditions (poorly kept facilities). Some raptors are more susceptible to this disease such as arctic birds (gyrs and snowy owls), goshawks, red-tailed hawks and golden eagles. Difficult to detect as symptoms are vague but alarming signs are depression and weight loss, vomiting and hoarseness or loss of voice. If diagnosed early, it can often be cured with antifungal treatment, though some birds may require surgery or prove beyond cure. However this disease is usually quite difficult to treat and almost always fatal. If any symptoms are suspect, please consult your vet immediately for diagnosis and ELISA test, which can at times already confirm the illness before symptoms ever appear.

To avoid this disease, it is important to maintain maximum hygiene and avoid stress. Humidity can be a problem as it can create an ideal home for this fungus. Giant hoods and transport boxes should also be cleaned and aired out regularly to prevent this disease.

Bumblefoot

Also known as ulcerative pododermatitis, this is a bacterial infection and inflammatory reaction on the feet of our raptors. This condition seems to be unknown in the wild and though there are many factors associated it with it (genes, bad circulation), the most common causes are poor hygiene, inadequate perches (which again contribute to bad circulation), and talons that need to be trimmed back plus cuts and grazes which could become infected. Treatment is difficult for this condition and there can be various phases from simple treatment with antibiotics through to operation. However this at times can be so severe as to cripple a raptor or even cause death.

Large, heavier birds, like red-tails could develop this condition and any bird that is not flown regularly and kept in the best sanitary conditions. It is better to avoid this condition than have to treat it though I must say that some cases can have hope. While in Spain, a young saker that I flew and worked with seemed to have a chronic case of bumblefoot and even though he had been operated and it had seemed to disappear, this condition came back to haunt and torment the poor bird, who seemed would do anything to not stand on his legs.

After a while, I decided to make a home-remedy; a face-cloth towel (the type that you can put your hand inside), an ice-pack and a little ribbon to tie the ends of the "bumblefoot pack" was all that was needed. I put the ice-pack inside the towel, thin enough to let the cold through but not so directly, tied the ends together (to not let the pack fall out) and put it underneath the poor and unsuspecting "Figo" who was "trying" to stand up on the screen-perch. I thought he would throw the pack right off, but he didn't. I began doing this a few times a week and the swelling seemed to go down in a short-while. What's more, "Figo" seemed to enjoy his moment of pampering and would only throw the pack on the floor to tell me when it was warm after a while (meaning he wanted another one!). He was up and flying in no time and even though like magic, the bumblefoot seemed to have disappeared. I continued this as a preventive treatment 3 times a week to also encourage blood circulation and provide a nice treat. I have since then recommended this helpful tip to many falconers and students and seems to work well, although severe cases may not be cured completely.

Parasites

We should always keep a close watch on the weight of our red-tail as any significant amount of weight loss could be a sign that it is suffering from parasites. At times, raptors can be seen to appear uneasy, moving their wings and with an unexplainable increased appetite. These can be early symptoms that could suggest that your raptor is suffering from a high level of parasites. Treatment is simple and quite efficient, however, before diagnose, we must make sure to always consult a vet and to make sure we have ruled out the possibility of stress that could be causing the weight-loss. This situation tends to be quite common when we get a new bird, as she has to get used to us and will be suffering from stress and also in the initial phases of manning/training.

Internal parasites

- Coccidiosis
- Flukes
- Worms (round worms, tape worms and capillaria)

While it is not usually common to see in captive bred birds, many healthy birds of prey will have parasites in the wild. For this reason, usually vets recommend worming a bird at the time of trapping followed by a follow-up treatment two weeks later, and then an annual check-up and worming at the end of the hunting season both to clear the system from anything picked up during the season as well as prepare for the moult. However while low levels are normal in wild birds of prey and can even go unnoticed, an increased number of these could cause serious health problems, starting with weight loss right through to weakening our bird. A weak raptor would then easily be susceptible to other illnesses and conditions.

To avoid elevated levels of parasites, one of the best ways of doing this is to always provide fresh food to our raptors and remove crop, head and intestines as these are important infection areas. We should never allow raptors to eat food that has gone off or that has been outside for quite a while.

External parasites
- Flat flies
- Feather lice
- Fleas
- Maggots
- Mites
- Ticks

These parasites may also affect our raptors and are quite common, such as the feather lice that can eat away their plumage. Again, we must examine our birds and their feathers regularly and look out for feathers that appear to be "eaten away" or "chewed". These parasites are quite contagious and can destroy the entire plumage of our birds. To avoid this, it is best to regularly spray our birds with a mild avian insecticide.

First Aid

Red-tails are pretty indestructible as far as birds of prey go but they also have their Achilles tendon. As surprising as it may seem, the most vulnerable parts of a red-tailed hawk are actually the skull (which is quite thin in order to reduce weight) and the neck, which is vertically quite strong but not so from side to side. Depending on the subspecies, the tarsi are also quite vulnerable and a squirrel or hare bite on them could even cripple a hawk.

If something does happen to our bird or we have reason to suspect she may be ill, we must always take her to the vet as soon as possible. However, if we are out hunting and far away from the vet, it would help to have a first aid kit ready for any emergency to help treat our bird on the way to the vet.

I also strongly recommend reading up on first aid and there are a few publications which are particularly helpful such as "Field First Aid for Birds of Prey" by British vet Neil Forbes, who also now teaches regular first-aid seminars in the United Kingdom to falconers.

In any case, I recommend having both at home and in the car, a small first aid kit made up of at least the following:

- A small clean blanket or towel, that could help us if our bird is wet against hypothermia and to also cast her or hold her down for examination.

- Tweezers, scissors or nail clippers which could be handy for removing thorns.

- Water or plant spray which can be used in general to clean the nasal openings when birds have a cold and may not breathe well as well as help to disinfect small wounds and cuts when out in the field. My favourite was an aloe vera spray which was sold by the IBR shop and was quite good for this purpose as well as for extra feather care. In any case, we should have at least a Betadine or iodine solution to treat these small injuries and gauze pads, cotton balls or paper towels.

- Electrolyte (or similar solution), which can be used together with crop syringe and tube in case of shock, although if this happens, I would recommend you get to your vet as quickly as possible.

I also suggest you put together a list of important phone numbers and tape them to your first aid kit or bag. Be sure to include your vet's phone number, several numbers for emergency 24h vet hospitals, animal poison information numbers and any other important numbers which you could need in an emergency.

The Red-tailed Hawk

A pair of red-tailed hawks in their nest at one of New York's most exclusive buildings, just in front of Woody Allen's apartment: their favourite place for dates and romantic encounters. Here we can see pigeon spikes for pest control, which have harmed this nest and their red-tails on several occasions- The nest, and all of its twigs are protected under the migratory bird treaty and removing the smallest twig could result in a heavy fine. However, to my surprise I was contacted in autumn of 2004 by Frederic Lilien and Lincoln Karim who were doing their best to make sure the law was indeed enforced as the nest had been removed on the basis that it was no longer an active nest (of course, red-tails don't breed out of season!) by certain people who seemed offended by it. I remember this incident well and luckily, after many efforts, the red-tails were not evicted from their home. Let's hope this situation never happens again and that we see many red-tails in the soaring high above Central Park.
Photograph by Lincoln Karim.

8. Red-tails in Love

The red-tailed hawk does not only arouse passions; it is itself a passionate bird which derives intense pleasure from each moment of its life, not only when it hunts its prey as the born hunter that it is. It also derives joy from the first rays of morning sunshine, from gliding through the sky in a way that makes us think that flying is as easy as dreaming. This bird captures our heart when we least expect it—much like our first love. In this way, it works its way into our hearts, and delights our innermost soul.

Love is also something real for these marvellous creatures and, like all other passions; it is experienced to the fullest. Red-tails fall in love and, provided that their partners do not die prematurely, they remain united throughout their lifetimes. A good example to follow. It is in Central Park, the largest park in New York City, where the most famous and most photographed red tail couple makes their home. This pair of red-tails is the object of fascination of thousands of birdwatchers, some of whom are out in the park at 4 or 5 in the morning to observe them before heading off to work, and some of whom sacrifice their lunch hour to catch a glimpse of them. Some have even taken jobs that allow them to view the birds more frequently during the day. This is a strange phenomenon: Many passers-by, upon seeing the crowd of photographers snapping pictures of the balcony of Woody Allen's apartment, assume that they are awaiting an appearance by the legendary director himself. But no. These red-tail devotees are instead waiting for Pale Male and his partner—the most photographed red-tails in history—to appear. One of these photographers is **Lincoln Karim**, who took the marvellous pictures of these two birds that he has graciously allowed me to publish in this book. It seems as if there is no movement of these two birds that he has not recorded, as if these two birds were his own children. He has in effect created a kind of "family album" with the pictures that he has taken during the past few years. He seems to be the person who is most intimately acquainted with these wonderful creatures, and each of his photographs seems to be a true work of art, reflecting the magic that these birds have bestowed upon Central Park and the delight that they have imparted to those whose hearts they have captured.

In Central Park, there is a register book that carries the observations of professional and amateur ornithologists about birds that they have observed in the park. This book is kept in the Loeb Boathouse, and is accessible to the public. Anyone can read it and add his or her own personal commentary about the birds or other fauna of the park. It is in this book, which was the brainchild of Sarah Eliot, that observations regarding red-tailed hawks were recorded that came to serve as the basis for the beautiful novel of Marie Winn, titled "***Red-tails in Love: A Wildlife Drama in Central Park.***" This book tells the incredible story of the two red tail lovers of Central Park.

The Red-tailed Hawk

This is the story of "Pale Male," which is the name that the male of the couple received on account of his pale colour, and of his companion "First Love." Red tails seem to have passed through the park during their migrations, with some 1,360 observations having been recorded over the years. They are most frequently observed during the month of November. However, these birds of prey had always been transient visitors and had never actually stayed in the park for any length of time—that is, until the arrival of Pale Male in 1991. This fact in itself is not only

Pigeon; quite usual in the diet of this family of red-tails and which on one occasion, was responsible for the death of "First Love". The first partner that Pale Male ever had. It is ironic that the death of one of the biggest predators of pigeons in the park should be caused by poison used to kill such pigeons. After the unfortunate incident, the use of poison for pigeons was banned, as the Buteo jamaicensis, its natural predator, is one of the best "pigeon-scaring" systems that can be used to get rid of them. Lincoln Karim, who took these wonderful photographs, told me that Pale Male caught most of these pigeons in flight, as it has been recorded on film and broadcasted by various TV programmes and news channels. This shows that the red-tails are also contributing with a good deed to the city and controlling what could otherwise become a real plague.

remarkable because it represents the first known instance of a red tail remaining in the park, but because it represents atypical behaviour for red tails as a species. As we say in the first chapter, red tails do not typically construct their nests near places frequented by human beings. However, Pale Male and First Love live their lives in close proximity to humans as if it were the most natural thing imaginable, seemingly undisturbed by the clutch of admirers that gathers to observe their every move—even during mating season. In the US, this is the first known case of red tails making their home in an urban environment.

These red tails seem to be entirely at home in Central Park, constructing their nest in buildings like the one seen in the photograph on an earlier page, very close to the apartment of Woody Allen, hunting and playing in the park and in the streets of New York, and even copulating in broad daylight, oblivious to the shouts of glee and the clicking of the cameras of their many admirers. This is all very surprising.

Pale Male and his family decided to settle in Central Park some years ago. Pale Male is now around 16 years old, and has been observed since his first moult in the park. The birds have found a home in the park—one that is quiet and that has plenty of food. The couple feed mainly on pigeons, rats, squirrels, and rabbits that are in the park. It is not surprising, given these circumstances that they have decided to stay put. Most of us would do the same thing if our needs were met so perfectly.

Red-tails In Love

The novel tells the story of red tails that share Central Park with the inhabitants of New York during the six years that Marie Winn spent in the park along with other nature lovers and amateur ornithologists—the so-called "regulars". The book focuses on the park, on nature, and on the red tail couple that has captured our hearts. Marie tells the story of her six years in the park—but she continues to go to the park, and plans to continue to do so indefinitely. We can go to Central Park ourselves, and begin to understand something of what she herself has experienced.

Everything began with the bird register located in the Loeb Boathouse and that is accessible to the public for both perusal and the recording of observations. This register was started by Sarah Elliot, and two of its primary contributors have been Tom Fiore and Norma Collin.

Scene One – Enter Pale Male

The name of the game in bird-watching is telling one species of bird from another. This one's a chickadee; that one's a nuthatch. But only by marking birds in some way, usually by attaching bands to their legs, can individual birds within a species be distinguished from each other. Generally, one robin looks like any other, or one downy woodpecker, or one red-tailed hawk.

The Red-tailed Hawk

Every so often it happens that a particular bird displays some feature that makes it recognizable as an individual. Sometimes there's an injury – a duck with a broken wing; it's simply a natural oddity of size or shape or colour. People may follow the course of such bird's lives in a way that would be impossible if they looked just like their species-mates.

Such a one was the red-tailed hawk that arrived in Central Park during my first winter as a Regular. He had a feature so distinctive he could always be identified – not just as a red-tailed hawk but as himself, a particular, individual bird. Whereas this species appears in field guides with a white breast, a broad band of streaks across the belly, and a darkish head, this particular red-tail was exceptionally light all over. His head was almost white. He had no belly-band to speak of – the breast and belly were white. He wasn't an albino, his eyes were too brown; just a very pale red-tail.

Tom Fiore saw him first, and reported his sighting in the Bird Register on November 10th:

"There is a very light-coloured, immature red-tailed hawk that has been eating a rat and also swooping a foot above shoveller ducks on the lake."

How did Tom know the light-coloured hawk was immature? Was it smaller?

It's a common misconception that all young birds are smaller than adults of their species. Though it is true for ducks, geese, and birds that leave the nest shortly after hatching, most songbirds and birds of prey, indeed, the majority of the avian order, cannot leave the nest until their flight feathers grow in. These birds are full adult size by the time they take their first flight.

It is the plumage of most immature birds, not their size that clearly distinguishes them from the adults. Young birds generally lack whatever bright colours adults of their species might display, making them less conspicuous during the vulnerable nestling and fledgling periods. Young robins have speckles on their breast rather than the characteristic solid brick red colouring of adult robins; juvenile male cardinals are buffy brown, not bright red, as their fathers are. Immature red-tailed hawks contradict their name: their tails are brown and won't turn red until they are two years old – breeding age. Moreover, the tails of immature red-tails are marked with distinct black bands. One look at the pale-coloured red-tail's tail revealed to Tom that this one was a kid.

For a while Tom was not sure whether the young hawk he'd sighted on November 10th was a male or a female. Among most birds of prey, males and females do not significantly differ in plumage or markings. Still, there is a way to tell the sexes apart: Female raptors are almost always bigger than males. Consequently, when a second and noticeably larger hawk arrived in the park a few months later, the sex of the pale-coloured bird became perfectly clear. The Regulars began to call him Pale Male.

He was still a browntail that spring, too young for love. Nevertheless and notwithstanding, when the female (this one with a bright red tail) showed up at the beginning of March, Pale Male courted and won her.[1]

8. Red-tails in Love

Both Marie Linn and Lincoln Karim participated in the movie about these magnificent red-tails that was directed by Frederic Lilien, and that is titled Pale Male: The Movie. It premiered in New York in 2002 and won the audience award for best documentary at the Brooklyn International Film Festival. The screenplay was written by Janet Hess, and features a number of celebrities who have become interested in the lives of Pale Male and his family.

Narrated by the actress Joanne Woodward, the story begins like a classic fairy tale.

"In the winter of 91, a stranger came to Central Park. Of all who flocked to the city, he was the first of his kind. He possessed a special power for, overnight, he transformed the heart of the city from something artificial into something real. The stranger was a red-tailed hawk. Whiter than most red-tails he was unlike any hawk in history. None had ever attempted what he would achieve, and his coming would change more than Central Park. It would awaken something all but forgotten in the many New Yorkers who followed his story. It all began when they named him 'Pale Male'.

This beautiful story has what appears, at least for now, to be a happy ending since, after all these years, the story still goes on... More than a decade has passed since Pale Male met "First Love". He is now with his fourth partner and is the father of 19 chicks, and is thus the founder of his own dynasty of red-tails—thus appearing to ensure that the magic will continue for many years to come.

Everyone has enjoyed this experience, especially small children. For this reason, I recommend that all of you who are not able to travel to New York yourselves visit the websites of Lincoln Karim, Marie Winn and the film (see references at the end of the book). On those websites, you can see more photos, and you can even order a DVD of the film. You can thus come to know the red tails that have not only captured the hearts of New Yorkers, but of the entire United States.

---

[1] *Excerpt from pages 43–45 of Marie Winn's "Red-tails in love: A Wildlife Drama in Central Park. Published with permission from the author for this book.*

The Red-tailed Hawk

Best dates for viewing migratory raptors over Central Park

	Sightings		Peak viewing period		
Species	**Earliest**	**Latest.**	**Sept**	**Oct**	**Nov**
Bald Eagle	25-Aug	13-Dic	18…25	5…12	
Osprey	15-Aug	6-Dic	1	15	
Golden Eagle	25-Sep	12-Dic			
Northern Harrier	18-Aug	14-Dic	15	20	
Northern Goshawk	15-Oct	14-Dic			
Broad winged hawk	15-Aug	22-Oct	10…25		
Red-tailed Hawk	29-Aug	15-Dic			10…25
Red-shouldered Hawk	20-Sep	14-Dic			10…20
American Kestrel	15-Aug	12-Nov	1	18	
Merlin	22-Aug	24-Nov	15	15	
Sharp-skinned Hawk	23-Aug	13-Dic	10	11	
Cooper's Hawk	23-Aug	8-Dic	10		10
Peregrine Falcon	27-Aug	13-Dic	15	10	
Rough-legged Hawk	4-Nov	13-Dic			
Turkey Vulture	15-Aug	10-Dic	25	10	
Black vulture	19-Sep	15-Dic			

Adapted from page 283 of Marie Winn's "Red-tails in love: A Wildlife Drama in Central Park, published with the author's permission for this book

8. Red-tails in Love

Pale Male observes the traffic of Fifth Avenue in New York (top photograph) and below, he flies over the same avenue. Photographs by Lincoln Karim.

The Red-tailed Hawk

Pale Male at sunset. Photograph by Lincoln Karim.

Pale Male on hunting flight. Photograph by Lincoln Karim.

8. Red-tails in Love

Bringing construction material to the nest (above) for the new baby Pale Male at sunset. Photographs by Lincoln Karim.

The Red-tailed Hawk

Adult red-tailed hawk. Drawing by Dr. Paul A. Johnsgard.

9. Useful Addresses

I have decided to include a list of useful websites for this section which could be of interest, both on red-tailed hawks and buteos in general as well as contributors to this book, and a small selection of useful addresses of websites, clubs, photographers, etc. Please visit the worldwide directory of my website at **www.yarakweb.com** for a more detailed and comprehensive list by country.

Websites

www.yarakweb.com

The world's unique multilingual falconry portal and fastest growing on-line raptor enthusiast community. News, weather, RSS feeds, articles, forum, chat, equipment making, tips, a worldwide directory and a falconry shop plus much more! All free, by falconers for falconers!

http://squirrel hawking.homestead.com/home.html

Website on squirrel hawking by falconers Gary Brewer (author of Buteos and Bushytails) and Manny Carrasco. Full of interesting photos, recommended articles, falconry toons, and information about their squirrel hawking DVDs. Highly recommended.

http://www.palemale.com

Website on the red-tails at Central Park; Pale Male and his family. Here you can find information, photographs and all you ever needed to know about them. You can also buy the movie on them (highly recommended). Photographs and website by Lincoln Karim with Frederic Lilien.

http://www.mariewinn.com

Author and journalist Marie Winn's website. Includes information on bird watching including the red-tails in Central Park and on her various books and publications, featuring "Red-tails in Love: A Wildlife Drama".

http://www.falconer.dk/

Morten Clausen's website with information on birds of prey, breeding and falconry in Denmark.

http://www.pbase.com/lorenzde

Dennis Lorenz's website. Dennis is both a falconer and photographer with a Bachelors Degree in Zoology/Ornithology at Michigan State University. He is currently studying his master's degree while he continues with professional photography and is a regular contributor of Yarakweb.

http://www.falconry.ca/

Roy Priest's website on falconry in Canada.

www.golden-eagle.org

Charlie Kaiser's website, photographer and falconer, full of interesting photos on red-tailed hawks as well as other raptors.

www.gsrenterprises.com

George Robertson's website which features information about his video on a Swainson's family: A hawk's nest. Highly recommended.

www.swainsonshawk.org

Friends of the Swainson's Hawk is dedicated to the survival and recovery of the California Sacramento/Central Valley population of Swainson's Hawk.

www.Benlongfalconry.co.uk

Falconer and equipment maker since the '70s, Ben Long has made quite a name in falconry and his name is synonym for quality and elegance in falconry products.

www.wildspirits.it

Claudia Panniello's website. Claudia and her husband Ageo live in Italy and make leather and all sorts of decorative elements with animals painted on them, as well as falconry gloves, like the beautiful falconry glove she made for Bandit (included in the equipment section on Chapter 4).

www.davidlprescottwildlifeart.com

Website of naturalist painter and photographer David L. Prescott. David has been featured artist at a number of shows and in magazine and newspaper editorials such as "The Falconer" magazine. His original paintings are in private collections throughout the world and have been exhibited at prestigious locations including the main falconry events in the United Kingdom.

Clubs, organizations and associations

International Association for Falconry and Conservation of Birds of Prey

Dedicated to the preservation of the ancient art of falconry, a hunting tradition defined as ' taking quarry in its natural state and habitat by means of trained birds of prey'. The IAF currently federates 63 falconry clubs from 48 countries.
www.i-a-f.org

Birdlife International

BirdLife International is a global Partnership of conservation organisations that strives to conserve birds, their habitats and global biodiversity, working with people towards sustainability in the use of natural resources. BirdLife Partners operate in over one hundred countries and territories worldwide.
www.birdlife.org

NORTH AMERICA

North American Falconer's Association (NAFA)

The North American Falconers Association (NAFA) is a group of dedicated individuals who share common views on the welfare of raptors (birds of prey) in nature and in their careful employment in the sport of falconry. NAFA welcomes you to the world of raptors and invites you to explore the ancient sport of falconry.
www.n-a-f-a.org

9. Useful Addresses

British Columbia Falconry Association
The BC Falconry Association is dedicated to protecting and extending the privileges of falconers, in the context that the sport of falconry represents both a responsible and conservative utilization of both wild and domestically bred raptors. http://bcfalconry.ca/

UNITED KINGDOM AND EUROPE

British Falconer's Club (BFC)
Founded in 1927 - The oldest and largest hawking club in the country. The British Falconers' Club is dedicated to the preservation of the ancient art of falconry and the conservation of birds of prey. In 2002 the club celebrated its 75th anniversary, as we continue to uphold the finest traditions of this, the noblest of all field sports.
www.britishfalconersclub.co.uk

British Hawking Association (BHA)
When the BHA was first formed it had two main objectives; firstly to create an environment where anyone wishing to become involved in falconry could learn from experienced falconers, and secondly, to bring together falconers from all walks of life to stand fast in the fight to keep falconry a legal and recognised sport.
www.bha.org

Asociación Española de Cetrería y Conservación de Aves Rapaces (AECCA)
Spanish association for falconry and conservation of birds of prey.
http://www.aecca.org

Nederlands Valkeniersverbond Adriaan Mollen
A Dutch falconry association.
http://www.adriaanmollen.com/

Association Belge de Fauconnerie "Club Marie de Bourgogne"
A Belgian falconry association.
http://users.belgacom.net/Mariae-Burgundiae/

Deutscher Falkenorden (DFO)
German falconry association.
http://www.falkenorden.de

Yarak Club di Falconeria
An Italian falconry association.
http://web.tiscalinet.it/yarak/

Association Nationale des Fauconniers et Autoursiers français
French falconry association.
http://www.anfa.net/

Schweizerische Falkner – Vereinigung
(Swiss Falconry Association)
http://www.falknerei.ch/

Osterreichischer Falknerbund
(Austrian Falconry Association)
http://www.falknerbund.com

Sokolarski Centar (Croatia)
http://www.sokolarskicentar.com/

Associacao Portugesa de Falcoaria
(Portuguese Falconry Association)
http://www.apfalcoaria.com/

The Red-tailed Hawk

Juvenile red-tailed hawk with adult feature on tail. Drawing by Dr. Paul A. Johnsgard.

10. Glossary

While falconry literature is varied and available in various languages, many valuable treatises, essays, articles and books that have laid the foundations for modern falconry today have been written in English. The language of falconry has evolved and grown since its earliest days to incorporate a whole new collection of modern terms and innovations. These are important contributions to an already rich terminology and must be clearly understood by falconers worldwide if they are to grasp new concepts and techniques.

For this reason, I have devoted much time to studying and learning the language of falconry and even wrote a book about it "***Ars Accipitraria: An essential multilingual dictionary for the practice of falconry and hawking***". My interest in the language of falconry was born both out of a personal interest in languages, as professionally I have worked as a qualified translator amongst other things but also due to being a falconer, as if one wishes to learn as much as possible about this noble art, one must go through a vast amount of literature, at times with very obscure terminology and in different languages.

I have therefore decided to include a small selection of the most relevant general terms in English for falconry as well as some very specific "red-tail/hawking" only terminology to help clear up any doubts that could arise from reading the pages inside this book or falconry publications in general.

For a more complete version of falconry terminology in six European languages (English, Spanish, Italian, French, Dutch and German), please refer to my book ***"Ars Accipitraria: An essential Multilingual Dictionary for the Practice of Falconry and Hawking"***, available through our falconry website shop at **www.yarakweb.com** or through our publishing website at **www.yarak.co.uk**.

.

General Falconry Glossary and specific "red-tail terms"

Accipiter	A short-winged hawk with short, rounded wings and a long tail, typically found in forests and woodland. This genus includes the short-wings of traditional falconry, such as the goshawk and the sparrow hawk amongst others, but also the Cooper's hawk.
Aerie	The nest of a bird of prey.
Albinism	This is a congenital absence of any pigmentation or coloration which also affects birds of prey, and seems to be more prominent in buteos, particularly in red-tailed hawks.
Anklets	Sometimes also called bracelets. The anklets are simply leather straps which go around the bird's leg. The hunting or field jesses (slit-less) are attached to this. These are recommended for hawks and owls and help avoid unnecessary accidents while flying through woodland areas or bushes like for example getting caught on a small branch.
Arm a bird	This means to put the jesses, bewits and bells on to a bird of prey.
Arm-brace	A support for the arm used by some falconers for carrying large birds on the fist, such as a golden eagle. Similar to a crutch.
Austringer	One who keeps, trains and hunts with short-winged hawks, also known as a shortwinger. This term was formerly used when referring to a person who kept and flew goshawks.
Aviary	The place where birds of prey are kept, whether tethered or free-lofted.
Aylmeri	An alternative to the traditional jesses, these leather anklets were designed by Major Guy Aylmeri and are made up of three parts: anklets or bracelets (fitted around the hawk's tarsus) with interchangeable jesses: mews jesses with a swivel slit (fitted whenever the hawk is tethered), and slit-less field jesses which are fitted whenever the hawk is flown free. Aylmeri jesses are strongly recommended for hawks or any bird that has tendencies to fly in dense woodland, trees or bushes and help avoid accidents.

10. Glossary

Bagged quarry — Live prey, disabled by falconer (i.e. feathers plucked from a pigeon to make it fly less) and let out freely for the hawk to chase. Normally used only when entering a hawk into a prey or when a natural quarry is very scarce, to ensure the hawk will get a flight. This is illegal in the UK and USA and many countries worldwide.

Bail (to bail) — When the squirrel sees no way out and jumps from one tree to another or to the ground, where the chase may continue.

Bait the lure — To tie the meat to the lure or garnish the lure. Usually on both sides, so no matter how the lure lands, the bird still gets his meat clean.

Bal-chatri — Also more commonly known as a BC, this is a trap often used by American falconers and banders, made up by a container and a series of slip-nooses (nylon) to entangle the feet of the raptor.

Ballooning — Helium-filled balloons now widely used around the world to haul the garnished lure high above the falconer. Balloons and kites are now becoming a very popular alternative to the more traditional falconry methods.

Banding — To determine hawk populations in the wild, many hawk observatories (i.e. Hawk Mountain) hold "banding programmes" through which wild hawks are caught, measured, marked and banded with an identification tag or ring. This is useful and helps determine migration trends as well as population numbers.

Barred; barring — Dark stripes that appear on the feathers of some birds of prey, sometimes as a feature of young adults (like barring on the breast feathers of some juveniles) or of a species or subspecies (barring on the tail of eastern red-tailed hawks).

Bate (to bate) — When a bird attempts to fly away from a perch or the fist while tethered or secured to either its perch or the falconer's glove. Something may have startled it or perhaps it has seen a possible prey or be impatient (in yarak) to be out flying or hunting.

Bechins — A few pieces of meat given to the hawk as a treat, especially after performing well.

Bells	Before the arrival of telemetry, bells were always used in falconry and played an important role alerting the falconer to the bird's location in the field. These bells were small and usually made out of brass, silver, nickel or stainless steel. Traditionally, two-toned bells, each with a different sound, would have been used for a bird. Even with telemetry, the use of bells is always recommended.
Bewit	Small leather strips which attach bells or transmitters to a bird's leg.
Bind (to bind)	When a bird grabs a quarry, lure, or the falconer's hand with its talons. Also, seizing quarry or lure with the feet in a tight, clamped-on hold.
Bird of Prey	A member of the Falconiformes and Strigiformes orders. Also known as a raptor.
Blink (to blink)	To refuse a quarry or slip.
Bolt	To move off suddenly, for short-winged hawks to fly straight from the fist at quarry. Also means to flush a rabbit from the bury, with the help of a ferret.
Bownet	Another commonly used trap with American falconers, which looks like a circle laying on the ground. As the red-tail approaches it to investigate the bait, it is sprung and a net is released over the trapped hawk.
Bowser	A bird that drinks too much water, perhaps due to dehydration or perhaps by malnutrition.
Braces	An alternative name for hood-braces. These are leather straps which when pulled, open and close the hood.
Branchers	An immature bird who can only jump from branch to branch in its nesting tree; this bird is learning how to use his wings but has not yet successfully flown and it is still fed by its parents.
Breaking in (to break in)	When our bird tears into the catch (usually after depluming it) and begins eating.

10. Glossary

Breastbone; keel
The breastbone of a bird of prey, often called "keel" by falconers. This is useful for indicating the bird's physical state. When a bird is too low or underweight, the keel feels sharp and bony. If however, it is not easy to feel the keel, then the bird is too fat and it is likely that the bird will not be hungry or interested in hunting.

Breeding (breeder)
To reproduce birds of prey in captivity, either traditionally, in chambers or through hacking. Red-tails should always be parent-reared.

Broadwinged
The common name for the species of Buteo or Parabuteo, typically known as the soaring hawks or also "windmasters".

Bushytail
Colloquial term used in USA to refer to squirrels, particularly with regards to squirrel hawking.

Buteo
Term used today in modern falconry to refer in general to hawks (more specifically buzzards), but correctly used to refer to genus of the same name.

Call off
Again, this is said of a falconer who calls the bird of prey back, either to the fist or to the lure.

Carriage
The carrying on the fist (glove) of a bird of prey during the early stages of the process of manning. The bird accompanies the falconer in his everyday activities, getting used to its surroundings, human activity, the area it will fly in, etc. In the old days, a way of doing this was feeding taking the bird to the busiest places (big kitchens of a large households or palaces), the town square, etc. to get her used to as much noise as possible and when finally used to it, get her to calmly eat on the fist, regardless of what was going on around her.

Casting
Ball of materials such as fur and bone which are not digested by the hawk and are thrown up. Usually called a pellet in owls, this term also refers to the natural materials (feathers, bone, grit, etc.) that falconers add to the hawk's food to help the bird cast. This is recommended regularly.

The Red-tailed Hawk

Carrying (to carry)	This expression is used to refer to a hawk that flies away from the falconer with its food, whether it is freshly caught quarry or food on the glove or tied to the lure. This bad habit can be wrongly encouraged on a bird of prey by the falconer in the early stages of training if he does not make in correctly and rewards the bird accordingly for its efforts. Remember, it is better to reward your hawk for his hard work and not fly a day or two, until the weight goes down, than to be impatient and try to speed up both the training and increase the number of prey the hawk can catch in a day.
Cere	The smooth, featherless, wax-like skin at the base of a hawk's beak into which the nostrils are pierced. Usually light yellow or green in young raptors, it can be a clear indicator of a bird's age and health. Also called the operculum.
Chick, one day old	Day-old chicks contain egg-yolk, which is full of nutrients and proteins, although also of cholesterol. This food can be an important part of their diet of raptors if not given too regularly and supplemented with calcium.
Condition	The physical state of a bird of prey (high or low), referring to its mental state (keen and eager to hunt, in yarak) and weight.
Coping	To trim or cut back an excessive growth of the beak or re-shape the bird's beak into its optimal form. This usually happens naturally, as the beak is used to tear and cut the food, but sometimes in captivity, this is not so, and the falconer must cope the beak a little.
Corkscrewing	Type of flight performed by the hawk when squirrel hawking, around a tree, as the squirrel and hawk are coming down from the tree.
Creance	A long string (no more than 20–30m) used when first training a bird, usually a new bird or one that cannot be trusted to fly free yet. Also used when accustoming a bird to flying in a new place. It prevents our hawk from flying away unexpectedly and losing her.

10. Glossary

Enseaming This term refers to getting a bird back in shape and to its proper flying weight after the moult or a period of rest (i.e. after an injury) by managing her food intake and getting her to drop the excess weight through diet and exercise. Basically means "getting rid of the excess fat" that a hawk has accumulated due to inactivity so that she can hunt again and give us an optimum response in the field.

Eyass A hawk taken from the nest: a nestling. When as traditionally, a bird is legally taken from the nest for falconry (please consult with local authorities), it is always described as an eyass, as opposed to grown birds trapped during their first passage in life.

Fat weight The total weight of a bird of prey, before it is trained and its weight is reduced to achieve flying or hunting weight.

Feed up (to) A bird of prey that has had its full ration of food. Normally the bird is "fed up" after performing well in training or on the field hunting. She will not be able to fly for a few days due to its weight being above flying weight (high). This term is also used when referring to a bird of prey that has just eaten and is not hungry anymore.

Ferreting This refers to hunting with ferrets (flushing prey out of cover) and birds of prey. It is common practice in the UK.

Fistbound A hawk that does not hunt wild quarry and seems to take the glove as its prey (also known as imprinted with glove). This can be a result of only introducing it to the lure as a method for retrieving your bird and is a bad habit encouraged by the falconer. Also known as "stuck" to the glove.

Fledge (fledglings) To abandon or leave the nest (though sometimes still under the care of the parents).

Flush (flush rabbits) To force game (such as rabbits) out of its cover so they can be hunted by our hawk by either walking towards them, beating the bushes or using dogs.

The Red-tailed Hawk

Flying weight	The highest weight possible at which the hawk is healthy enough to fly and hunt, yet sufficiently hungry to respond to the falconer's control and to any quarry that is flushed
Follow on, Following	This is usually said of hawks, when following the falconer from perch to perch in anticipation of quarry; however both Harris hawks and red-tails are great enthusiasts of this technique. A form of still-hunting.
Footing, to foot	When a hawk grabs to a lure, quarry or even falconer's hand using her feet to subdue or kill it, as hawks kill through pressure.
Fret-marks	Small, whitish marks found feathers, usually the result of stress or a bad diet while the new feathers are developing (during the moult).
Free-lofting	When a bird is free-lofted this means that she is allowed the full roam of her mews without being tethered. Although this can be recommended for most birds used in hawking (it can help protect the tail feathers avoiding constant bating), some birds do not acclimate to this well and some situations are not set up for this to be safe.
Giant Hood	Ventilated box used to transport hawks in complete darkness. This is called a "giant hood" as the effect is similar to that of a hood, if the bird were wearing it.
Gorge/ full crop	A complete ration of food.
Hackles	This term refers to the feathers on a bird's head, which are raised, when the bird feels threatened or angry, or when defending its prey, like red-tails and eagles.
Haggard	A wild adult hawk, of more than 12 months of age with mature plumage.
Hard Penned	This term was traditionally used when referring to a eyass whose feathers were finally fully-grown but is also used nowadays to refer to a bird of prey whose feathers have finished growing and has completed the moult.

10. Glossary

Hood	Used to keep the bird calm and reduce stress that could result from its surroundings usually with falcons as broadwings dislike it and are actually calmer when unhooded.
Hood-shy	A bad habit encouraged by a falconer who is not expert enough to know how to hood the bird quickly and properly. The bird becomes afraid of the hood and tries to avoid it.
Hybrid	A raptor bred and produced as a result of artificial insemination that contains genes of different species or subspecies (i.e. peregrine x lanner). Popular hybrids include gyr x sacre, gyr x peregrine and perlins but also hawks such as ferrutails.
Imping (to imp)	Cutting a broken or damaged feather and replacing it with an undamaged feather. The shaft of the bird's broken feather is cut, and the feather is trimmed to the right length. Then the shaft of the replacement feather is glued to the shaft of the broken feather on the bird (flexible glue). When your bird begins to moult, gather all the feathers dropped, select the best ones, and keep them, just in case, for future imping needs.
Imprint / social imprint	A bird raised by humans and not by other raptors; the bird will also tend to identify with humans, and will in general be a "screamer" (screaming imprint) for both food and attention. In general and save a few exceptions such as owls and birds of prey destined for breeding and hybridation purposes, birds of prey should not be imprinted if they are to be flown. This can be a particularly dangerous in the case of larger birds such as goshawks, red-tails and eagles. Generally, imprinted birds can never be released into the wild and will not recognise other birds as part of their species, as they have been reared by humans.
Intermewed	A bird of prey that has completed at least the moult in captivity.
Jesses (leather)	Traditionally made of dog leather, these are leather strips which go through the anklets so the falconer can hold the bird of attach the leash. Modern jesses are of many types of material including parachute cord and various braids.

The Red-tailed Hawk

Jump-ups	The exercise of jumping up to the fist from the ground or perch/block repeatedly for building up the muscle mass of birds of prey.
Laddering (up the tree)	When a squirrel and red-tail slowly climb a tree; the squirrel is trying to get away and the red-tail is trying to get ahead of the squirrel. This is used to refer to the movement of the red-tail (flying or jumping from branch to branch to gain height and get above the squirrel).
Lamping	The hunt of rabbit or hares with short or broadwinged hawks after dark, using torches or lights.
Leash	Traditionally made of leather, with a knot or button at one end, is what secures the bird to the perch or falconer's glove. Modern leashes are made of many materials those used for hiking such as nylon.
Longwings	This term is used for referring to falcons.
Lure	A weighted fake quarry used to train a raptor. A feathered lure with wings is used for training birds of prey into feathered quarry, while for entering into rabbits or furred quarry, the rabbit or dragged lure with the fur of a rabbit is preferred.
Making in	When a falconer approaches a hawk or falcon that has caught prey or has caught the lure. The falconer must come as quickly as he can in case his bird may need assistance but also to make sure the bird does not leave, as after having eaten and losing hunger, it could just fly away.
Manning	To tame a hawk or make a hawk or bird of prey. This is one of the first stages before its training and requires for the bird to be used to the presence of the falconer and be comfortable eating on the fist in any surroundings.
Mantling, to mantle over	Mantling is the action of stretching out the wings to hide food; a hawk will deck its feathers or plumage and attempt to cover a kill with its wings. This is common in many raptors, especially red-tailed hawks who seem to behave in a similar way to eagles, with their hackles raised and triumphantly standing their ground, warning us to not get too close. It is also believed to be deep within a bird's instinct as in the wild, they are often robbed of their freshly-caught preys.

10. Glossary

Meatball	The nest of a squirrel.
Mews	An enclosure or facility for keeping birds of prey; the hawk house.
Moult, (to moult)	This is the process of shedding old and growing new feathers which naturally occurs once a year. During this time, it is important to give our birds of prey food rich in nutrients (mice and rats are excellent for this) together with vitamin supplements and disturb them as little as possible, avoiding any situation which could produce them stress at all times.
Passage	A wild caught hawk on migration in immature plumage, before 12 months of age; an immature wild bird. This term can also take on the meaning of referring to a bird as being of wild origin, as is the case with many of today's more exotic falconry birds such as Harris hawks, Cooper's and red-tails or exotic eagles. These originally came from the wild but have been flown for several years as an adult bird.
Pick-up piece	A piece of meat with which to pick up a hawk; like for example a pigeon wing or piece of chicken, held in the fist used to tempt a hawk to step off a kill or the lure and on to the fist. This is usually a piece of meat of better quality than what they have caught in order to lure them off.
Preen / preening	When a bird spreads oil from the uropygial gland over the feathers and body through preening actions using the beak, usually after having a bath and while sunning.
Shortwings, Short-Winged Hawk	A true hawk of the Accipiter genus, for example a sparrow hawk or goshawk.
Soar Hawking	To glide or circle at a certain height, using thermals. This hunting technique is often used by broadwings, such as red-tailed hawks, who can achieve beautiful and impressive dives at great speeds, just like peregrine falcons.
Squirrel chaps	These are a type of leather anklets which cover part of the tarsus of a hawk and help protect it from bites or wounds. These are used in the USA in squirrel hawking and can also be a great aid when hunting hares or rabbits, as a bite by any of these can cripple a hawk for life.

The Red-tailed Hawk

Squirrel hawking — A popular practice in modern falconry where squirrels are hunted by hawks, usually red-tailed hawks but lately also female Harris hawks.

Sticky-footed — A hawk that has a tendency to strike-out at the falconer with her feet. Also used when referring to a bird with the bad habit of binding to the glove (also known as fist-bound).

Stoop, stooping — Hunting technique typical of falcons but also used by broad-wings where a bird of prey flies high in the sky, folds her wings back and drops downwards with great speed towards a prey or the lure.

Tiercel — A male raptor. The term came from the size of the male raptor, which is typically one-third smaller than the female.

Tiring — A tough piece of meat with bone and feather/fur (such as a pigeon wing removed at the shoulder or rabbit foreleg with little meat) given to our hawk for conditioning the beak and exercising the neck and back muscles but also when getting our bird used to our fist.

Upland game — Certain gallinaceous birds that can be taken as prey by hawks such as the red-tailed hawk and include pheasant, grouse, quail, partridge, etc.

Upwind — As opposed to flying downwind, flying against the wind. This can be helpful when exercising our birds and building up muscle tone.

Waiting on — When a bird of prey circles above a falconer waiting for game to be flushed (either by the falconer himself or through the aid of dogs) and soars either circling or hanging on the wind above the falconer, waiting for the game. When the game appears or is flushed, the bird will then stoop at the quarry.

Washed meat — Meat that has been soaked in water and lacks most of its nutrients. This can be helpful, if use for short periods of time, when enseaming our hawk or trying to reduce her weight in general.

10. Glossary

Watch / wake A medieval practice of keeping the freshly captured hawk on the gauntlet/glove for the first 24 hour or more, usually days at a time. The bird is present at all times in the falconer's daily activities and is never put down. Slowly, sleep and hunger win and the bird gives in and accepts food. This is the beginning of manning of the bird. Once she has accepted the food, the falconer will be able to work with her and eventually train her. This practice is still very much used in traditional falconry today by expert falconers.

Weathered To take a bird outside into the open air, usually after a good hunt or training exercise. This is generally done in a weathering lawn or enclosure to protect the bird from any other raptors, dogs, or cats while they bathe or drink.

Wetting a hawk The old practice of sprinkling or spraying a hawk with water to calm it down. This was done with hawks, such as goshawks who would persistently bate off the fist and also works well nowadays with " hyperactive" red-tails.

Wild-caught A bird of prey taken from the wild. This is still legal today in some countries such as the USA. Please consult with your local authorities.

Yarak This term was usually reserved for accipiters but in modern times, it is generally applied to all raptors used in falconry. An Eastern term, possibly of Persian origins and derived from the word "yaraki" meaning strength, power and ability. A bird in yarak is a bird in an optimum state of health and solely focused on the hunt; both body and mind. She is sufficiently hungry but not too weak to perform with success in the field.

The Red-tailed Hawk

"Bandit" at the end of his first moult. Photo by the author.

11. Appendix

Suggested diet for a red-tailed hawk (weekly)

The following sample diet is meant as a guide for beginners, taking into account the food section in Chapter 1 first, as you must always make sure you provide a balanced an healthy diet that is as similar as possible to what your bird would eat in the wild.

Casting material (feather, skin, bone, etc) should be fed to all falconry birds regularly at least a couple of times a week.

The food provided should be as fresh as possible or defrosted (naturally, do not use a microwave or any heating method to "defrost" food. This includes soaking the meat in warm/hot water as it will lose most of its nutritional value in this way and become "washed meat").

It is important to weigh your bird daily, always before feeding and that you feed/fly it at the same time approximately. It is also equally important to never reduce the weight of any raptor too quickly; this must be done gradually as it could harm your bird's health.

This sample diet is just meant as a quick guide for beginners and will need to be adjusted depending on the sex of your bird, season of the year as well as weather, current weight as well as other taking into account other factors such as fitness of the bird (muscle mass). It should be as varied as possible as it should not eat the same foods all the time. In any case, since many students always enquire about such "sample" diets, I have decided to include it as a small suggestion in my book.

Monday
Half a quail, fed with its feathers, bones (crushed please!) for the casting.

Tuesday
2 medium-sized mice.

Wednesday
¼ quail and 1 day-old chick (defrosted naturally).

Thursday
2 medium-sized mice and half a day-old chick (including the yolk).

Friday
1 pigeon/dove breast with its feathers.

Saturday
Crushed chicken wing with vitamin supplement (see food section, Chapter 1).

Sunday
Fasting. Today it would not eat so it is in flying weight for tomorrow, unless the weather is very cold. In that case, we could feed it 1 mouse or if we don't have any, half a day-old chick without the yolk.

The Red-tailed Hawk

Note:
Weight must always be lowered gradually. Red-tails will take time to get rid of their fat reserves but it is always much better for their health to do this slowly, than to be in a hurry. The above diet is meant as a sample diet and can vary for individuals and also according to sex, weather and other factors but was the diet I followed with my red-tail 3 or 4 weeks after finishing the moult to gradually lower his weight and get him flying again for the next season. Also, never lower a bird's weight when she is still moulting! Wait until she has completely finished the moult to do so, or her feathers could be damaged.

11. Appendix

Daily Log Book

Name of Bird:		Month:			Week:	

Day	Weather Conditions	Temp.	Weight	Slip/Training	Food	Comments
1						
2						
3						
4						
5						
6						
7						

Conversions

As falconers reading this book will be using both imperial and metric systems, I have decided to include a quick help-guide as a fast reference for converting these values. You should maybe photocopy these and have them handy.

	Multiply by	Value
MEASURE	**LENGTH**	
1 inch	* 25.4	millimetres
1 foot	* 30.48	centimetres
1 yard	* 0.914	metres
1 mile	* 1.609	kilometres
millimetres	* 0.039	inches
centimetres	* 0.39	inches
metres	* 1.094	inches
kilometres	* 0.62	miles
MASS & WEIGHT		
ounces – oz	* 28.35	grams
pounds – lb	* 0.453	kilograms
grams – g	* 0.035	ounces
kilograms – Kg	* 2.2	pounds
TEMPERATURE		
Fahrenheit	-32*5/9=	Celsius
Celsius	*9/5+32=	Fahrenheit

11. Appendix

Pattern for Squirrel Chaps

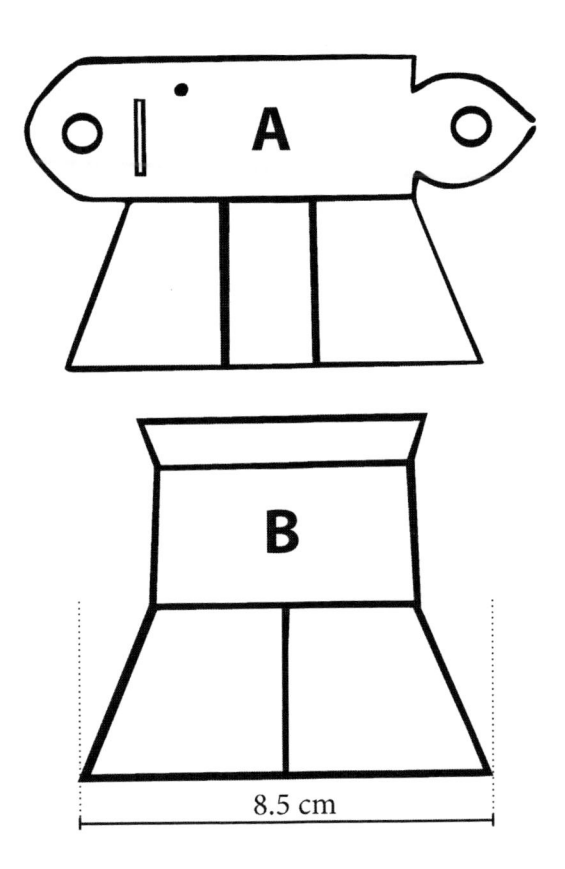

8.5 cm

Trace the outlines of A and B onto your piece of leather.
Cut out the pieces, following the outer edges.
Cut upwards on the two vertical lines in A, as far as the horizontal line below the "A"
Cut upwards on the single vertical line in B, as far as the horizontal line below the "B"
Glue the two parts together with A on top. (B acts as a reinforcement for A)

The Red-tailed Hawk

Patterns for Jesses & Anklet

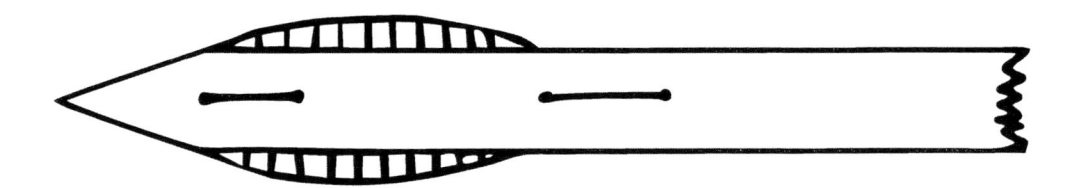

*Pattern for making traditional jesses.
The length will be approximately of 30 cm (12 in) depending on the bird of prey.*

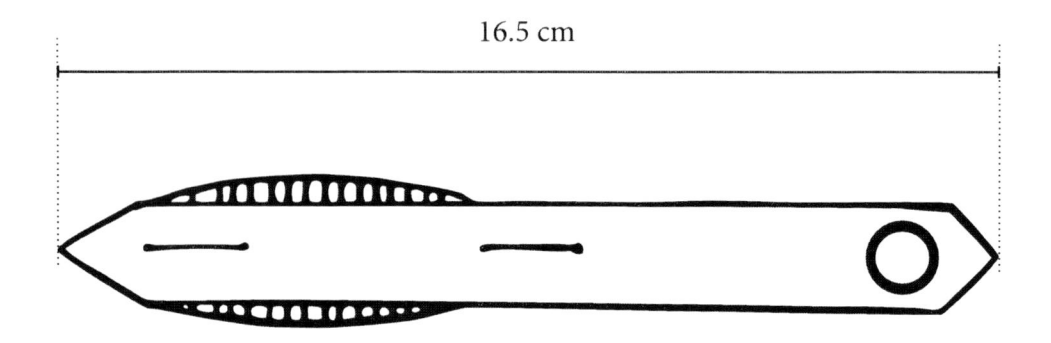

16.5 cm

*Pattern for false Aylmeri.
This is placed around the tarsi in the same way as traditional jesses,
save that these are shorter and end in an eyelet, just like the Aylmeri,
through which we will thread the real jess.*

Pattern for Aylmeri anklet, to scale.

11. Appendix

The Red-tailed Hawk

*Picture of male red-tailed hawk "Sid" after hunting a bunny in England.
Photo by Dave Hayter.*

12. Bibliography

Buteo jamaicensis

- **Austing, Ronald G.** *The World of the Redtailed Hawk*, (Living World Books). John K. Terres, editor. B.j. Lippincott Company, first edition: Philadelphia / New York 1964
- **Brewer, Gary L.** *Buteos and Bushytails.* GLB Publications, Texas: 1995
- **Butler, Daniel.** *Sharing the Seasons with a Hawk: The Redtail.* Jonathan Cape, 1994. ISBN 0-224-03867-2
- **McGranaghan, Liam J.** *The Redtailed Hawk. A complete guide to training and hunting North America's most versatile game hawk.* Third edition, 2001
- **Oakes, William C.** *The Falconer's Apprentice: A guide to training the Passage Red-Tailed Hawk.* Eaglewing Publishing, third edition: 2001 (Connecticut, USA). ISBN: 1-885054-03-3
- **Preston, C.R. & R.D. Beane.** *Red-tailed Hawk (Buteo Jamaicensis).* Birds of North America, No. 52 1993. (A Poole & F. Gill, editors). Philadelphia: The Academy of Natural Sciences; Washington, DC: The American Ornithologists Union.
- **Winn, Marie.** *Red-tails in love – A wildlife Drama in Central Park.* Pantheon Books, New York: 1988. ISBN: 0-679-43997-8

Other references

The following works contain detailed information on the Buteo jamaicensis and also on other North American birds of prey mentioned in this book.

- **Clark, William S. & Brian K. Wheeler.** *The Peterson Field Guide to Hawks of North America.* Houghton Mifflin; Second edition, 2001
- **Dunne, Pete and David Sibley and Clay Sutton.** *Hawks in Flight.* Houghton Mifflin Company, Boston: 1988. ISBN: 0-395-51022-8
- **Durman-Walters, Diana.** *The Modern Falconer: Training, Hawking & Breeding.* Swan Hill Press: 1994. ISBN: 1-85310-368-3
- **Garrido, Orlando H. Y Arturo Kirkconnell.** *Field Guide to the Birds of Cuba.* Cornell Univerisity Press: 2000. ISBN: 0-8014-8631-9
- **Heintzelman, Donald S.** *Hawks and Owls of North America.* Universe Books, New York: 1979
- **Hollinshead, Martin.** *The Complete Rabbit and Hare Hawk.* The Fernhill Press: 1999
- **Johnsgard, Paul A.** *Hawks, Eagles and Falcons of North America: Biology and Natural History.* Smithsonian Institution Press: 1990 Washington (USA) & London (UK). ISBN: 1-56098-946-7

- **Newton, Ian y Penny Olsen.** *Aves de Presa. Colección Materia Viva.* Encuentro editorial S.A., Barcelona: 1993. ISBN: 84-01-31485-2
- **Pareja-Obregon de los Reyes, Manuel Diego.** *Cetrería y Aves de Presa. Un duende de nombre Gavilán.* Cairel Ediciones 1997. ISBN: 84-85707-30-3
- **Peterson, Roger Tory y Edward L. Chalif.** *Peterson Field Guides: Mexican Birds.* Houghton Mifflin: 1973. ISBN: 0-395-97514-X.
- **Raffaele, Herbert & James Wiley, Orlando Garrido, Allan Keith, Janis Raffaele.** *Princeton Field Guides: Birds of the West Indies.* Princeton Univerisity Press: 2003. ISBN: 0-691-11319-X.
- **Snyder, Noel y Helen.** *Birds of Prey. Natural History and Conservation of American Raptors.* Voyageur Press: 1991. ISBN: 0-89658-131-4
- **Walker, Adrian.** *The Encyclopaedia of Falconry.* Swan-Hill Press, United Kingdom: 1999. ISBN: 1-85310-997-5
- **Wheeler, Brian K.** *The Wheeler Guide. Raptors of Western North America.* Princeton University Press: 2003. ISBN: 0-691-115-99-0
- **Wheeler, Brian K.** *The Wheeler Guide. Raptors of Western North America.* Princeton University Press 2003. ISBN: 0-691-115-99-0

Articles, publications and studies

- **Bildstein, Keith L.** *Behavioural Ecology of Red-Tailed Hawks (Buteo jamaicensis, Rough-Legged Hawks)* September 1987. ISBN: 0867271027
- **Brewer, Gary L. Redtail.** *Workhorse of Modern Falconry.* (Included in both the Spanish and English editions of this book with permission from its author, Gary Brewer). http://squirrel_hawking.homestead.com/workhorse.html
- **Brewer, Gary L.** *Getting the Most from your Redtail. Part one: Weight Management.* http://squirrel hawking.homestead.com/weight.html
- **Carrasco, Manuel.** In the Talons of Falconry. http://squirrel hawking.homestead.com/article.html
- **Faueaux, Mike & San Nottleman.** *Albino Red-tails.* American Falconry. Pages 30–34. Volume 9, December 1997
- **Gwiazdzinski, Jim y Geoff Dennis.** *Squirrelling in the USA, Rhode Island to be more specific.* International Falconer. Nº 13. May 2002
- **Hollinshead, Martin.** *Hill Hawk.* International Falconer. Nº 7. November 2000
- **Kaiser, Charlie.** *A Thousand Miles to Modoc.* Pages 42–48. Volume 13, December 1998. (Included in this book with permission from its author)

- **Sandstrom, Bruce.** *The Making of a Red-tailed Duck Hawk.* American Falconry. Pages 14–20. June 2002. Volume 27

Videos, films and documentaries

- **Attenborough, David.** *The Redtail. Life of a Hawk.*

  Excellent documentary on the life of red-tailed hawks in the wild.

- **Brewer, Gary.** *Introduction to Squirrel hawking I.*

  Video by Gary Brewer, author of "Buteos and Bushytails" on how to get started with a red-tailed in squirrel hawking. (This can be purchased in Northwoods).

- **Carrasco, Manny & Gary Brewer.** *Introduction to Squirrel hawking II.*

  Second video on squirrel hawking together with Manny Carrasco, friend of Gary Brewer and master falconer. Available on video and DVD through the author's website and also in Northwoods.

- **Frederic Lilien.** *Pale-male the Movie.* (available through Lincoln Karim's website at **www.palemale.com)**

  Film documentary on the red-tails that live in Central Park, new York. Really special film and a must for red-tail lovers, based on the love story written by Marie Winn in her book "Red-tails in Love".

- **Robertson, George**. *A Hawk's Nest.* (Can be purchased at http://www.gsrenterprises.com).

  A heart-warming story. This 59 minute independent video production was produced with digital video and audio equipment and contains a rich and exciting sound track of original music. It faithfully relates the true adventures of Junior, a newborn Swainson's Hawk, and follows his harrowing progress in a world fraught with genuine danger. These beautiful hawks face an uncertain future due to loss of critical habitat and the unfettered use of deadly insecticides in some of the southern countries to which they migrate every winter.

- **Kevin Byrd & Heath Garner** *Introduction to B.C. Trapping of Red-tailed Hawks & American Kestrels.* (Big Byrd Productions, USA).

  An informative video on the legal practice (USA) of trapping red-tails and American kestrels for falconry.

- **Timmons, Scott.** *California Hawking Club Meet in Yuba City 1999.* (Can purchase this also through Northwoods).

  This video shows the falconry meet at Yuba City, California. Here we can see many falconers catching rabbit and ducks in the yearly sky trials. Includes soar hawking flights of ferruginous and red-tailed hawks. An excellent video and one of my favourites.

About the author

Beatriz Elena Candil Garcia: Born in Madrid, Spain, in 1976, member of the Spanish Association for Falconry and Conservation of Birds of Prey (AECCA), appointed honorary member of the Mexican falconry association "Cetreros del Desierto de Coahuila" and member of various professional translation associations, Beatriz has studied political science (St. Louis University – USA), law (Universidad Complutense – Spain) and translation (both at CLUNY-ISEIT – University of Paris and London City University).

The author, now married and expecting her first child, has lived and travelled extensively all over the world and currently resides in both the United Kingdom and Spain. Beatriz is trilingual in English, Spanish and Italian, and possesses an excellent understanding of French in addition to basic German and Dutch.

Her interest for all animals was apparent at a very early age and during her first years at university, she volunteered in a shark research organization (PADI qualified diver) where she was later appointed as member of the board. Beatriz was also involved professionally in falconry while living in Spain since 2001, working as a falconer in bird control, training and successfully breeding birds of prey while also teaching falconry. She also has a weak spot for cats.

Beatriz has published several books and articles on falconry in various languages. Her first publication, *El Gran Desconocido: **The Red-tailed Hawk***, published in Spanish in 2004, won the award for Best Digital Book at the British Book Design & Production Awards.

A follow-up to this first publication was the unique multilingual falconry dictionary in six languages, "**Ars Accipitraria**: *An Essential Dictionary for the Practice of Falconry and Hawking*" published in 2007. With over 280 pages, it is a full and comprehensive study of the phrases and terms used in falconry. Covering six different languages – English, Spanish, Italian, Dutch, German and French – it is the only book of its kind.

Due to popular demand, Beatriz has now completed her translation into English, of an updated version of *El Gran Desconocido: **The Red-tailed Hawk**.*

Beatriz has also co-founded two new ventures with Arjen Hartman – her husband, and co-author of **Ars Accipitraria** – **Yarak Publishing** and **Yarakweb.com** the world's only multilingual falconry portal. Together they aim to unite falconry and falconers worldwide through future publications and projects.

The Yarak portal

As information about falconry on the Internet is either too cluttered, only available in the local language, or else not compelling enough, a unique falconry portal seemed to be the next logical step for Beatriz Candil Garcia & Arjen Hartman, authors of the multilingual falconry dictionary '*Ars Accipitraria*'. With Arjen's firm roots in Internet and the experienced author Beatriz combining their multilingual skills, the unique and exclusive falconry portal 'Yarak' was born in early 2007. By simply typing in **www.yarakweb.com** the user will be taken to *the* place where the modern falconer and general raptor enthusiast can communicate and find any information related to falconry, which will subsequently build awareness of falconry and worldwide conservation of birds of prey.

The English-Spanish (soon with more added languages) Yarak-portal is embedded in web 2.0 technology designed to facilitate the biggest falconry community on the planet. Via its chat rooms, forums, personal user pages and many articles of well known authors, falconers from all over the world can share and learn from each other.

In addition, Yarak will also be available via Mobile Internet and facilitating several handy tools that the modern falconer needs, such as a lost & found database, weather forecasts, the falconry dictionary, international raptor index, a powerful *falconry* search engine and a shop for books and quality falconry goods. Or one can simply relax by talking to fellow falconer friends and upload pictures & videos. So, whether you are behind your desk using a PC or out in the fields using your mobile phone to connect to the Internet, Yarak will be there.

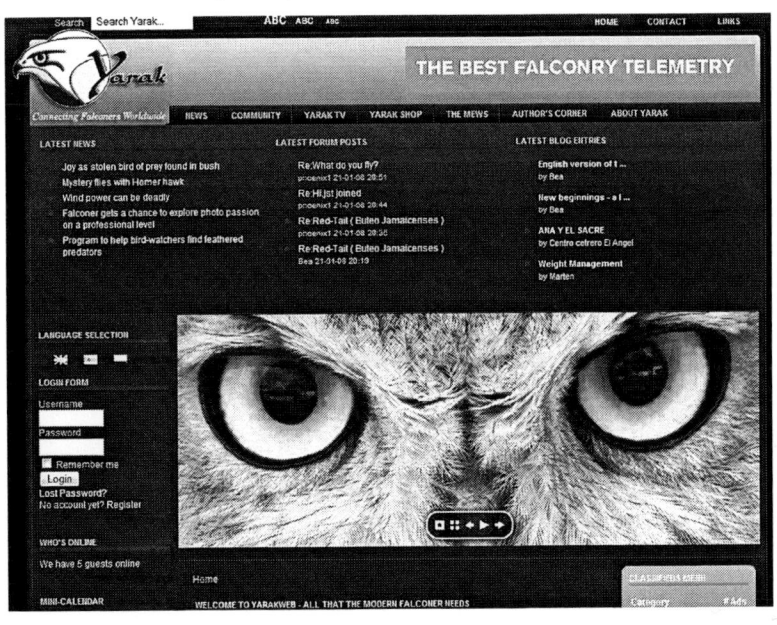

You can visit the world's biggest falconry portal at:
http://www.yarakweb.com *or* **www.yarak.mobi** *from your mobile phone*

Acknowledgements

First of all I would like to thank Bandit, a little male red-tail hawk I once had, for sharing his time with me and for inspiring me to write and publish this book in Spanish, back in 2004.

I would also like to thank my partner and husband, Arjen Hartman who has been my constant source of inspiration the last few years in all my projects and especially this book. I needed a final push, a lot of hard work and many sleepless nights to find the time to produce this English version and Arjen made it happen. Arjen has also contributed to this book with photographs and valuable information on falconry in the Netherlands.

A little big thank you also goes to unborn baby son, who made sure that I had many sleepless hours to work on this project and for whom I wanted to give this book to as a first gift and introduction to the red-tail. I finally managed to finish it before his birth which should be at the time of publication of this book, so I will indeed be able to read it to him in time and show him what we worked on so hard.

I would also like to thank my parents and my brother Carlos and encourage them to finally read one of my books, hopefully they will finally understand me a little more.

Last but not least, I would like to thank again David Cronin and Sally Cronin, for the second-time around, for their effort and dedication on this book, now finally in English, as well as their help in other projects and long-lasting friendship.

I would also like to thank Robert Forstag for his help in editing and translating this book with me.

Finally, I would like to thank the readers of this book for taking an interest and hopefully giving the red-tail the chance it deserves. When I first wrote this book a few years ago, I did so with much love for the red-tail, as a tribute, and also wishfully thinking that it would serve the purpose of attracting an interest to this magnificent yet quite overlooked bird in the falconry world. A few years have passed since then, but often, I catch myself day-dreaming, staring at the blue sky, hoping to find a little red-tail soaring beautifully up above me. I cannot help but smile and remember the good times I shared with my male red-tail; during many moments of my life, he had been a source of inspiration and happiness. In the same way, I sincerely hope that this book will provide you with additional knowledge and moments of joy, but most of all that it also serves to inspire you to embark on a new adventure and get to know the red-tailed hawk…

"The old shepherd was right: The only solution was to forget a part of uncertainty and create a new history for oneself."

Paulo Coelho - *The Fifth Mountain*

With best wishes for the New Year and hunting seasons to come!

Beatriz E. Candil García, 2008.

I would also like to give a big thank you to all the **collaborators / contributors** and acknowledgements to this book:

North America / South America

Brad S. Silfies

Brian K. Wheeler

Charlie Kaiser

Dr. Keith L. Bildstein, Ph. D, Mary Linkevitch (Hawk Mountain Sanctuary, Pennsylvania)

Frederic Lilien

Gary Brewer

Geoff Dennis

George Robertson

Jim Gwiazdzinski

Jonathan White in Florida (author of the antique-looking red-tailed hawk map on the inside cover)

Juan Manuel Iglesias (Puerto Rico)

Judith Lamare (Friends of the Swainson's Hawk)

Lincoln Karim

Manny Carrasco

Marie Winn

Paul A. Johnsgard

Raul Ducoing Arjona (Mexico)

Roy Priest (Canada)

Vicente, and Carolyn from Puerto Rico for her kind words and long-distance friendship"

Wendy Perrone

Europe

Spain

Mario Carabias

José Maria Cabrera, "Beta" and the team at Safari Madrid

Francisco "Pancho" Solano & family

José Manuel Gamito, Serafin Serrano & Raul Serrano and the Nucleo Zoologico Indalo

Ricardo & Tomás Brugarolas;

United Kingdom

Ben Long and his lovely wife Olga,

Craig and Jane Thomas

Edward and Liz Hopkins (Gwent Hawking Club)

Jeff Bower

Paul Beecroft, José Souto and the British Hawking Association (BHA)

Rest of Europe

Claudia Panniello for her beautiful drawings and most of all her friendship

Dennis Lorenz (Switzerland)

Fernando Manuel Flores (Portugal)

Marco Calistri, Luca Castiglioni and Andrea Brusa (Italy)

Morten Clausen (Denmark)

Wilfred Berendsen (The Netherlands)

Ars Accipitraria
an invaluable addition to any falconer's library...

Yarak Publishing is proud to announce the release of *Ars Accipitraria: An Essential Multilingual Dictionary for the Practice of Hawking and Falconry.* This book is without doubt the most comprehensive and up to date gathering of vocabulary on the noble art of falconry and birds of prey ever assembled.

Ars Accipitraria is the first work to attempt a compilation of terms that make up the international language of falconry, from medieval times to present day, aiming to make this vocabulary readily available to falconers and raptor enthusiasts worldwide. This 298-page multilingual dictionary represents the first multilingual compilation of its kind and contains circa 5,000 terms. It includes an English glossary of all the commonly used terms (both modern-day and archaic) and also provides brief but detailed explanations of the terms, often with tips from the authors, and translations of each term into S*panish, Italian, Dutch, German and French.* Ars Accipitraria also contains a unique multilingual index at the end of the book that has been carefully organised by language, making the book ideal not only for English users, but also for foreign readers.

Well organized, clear and easy to read, this valuable translation tool is a must-have for all raptor enthusiasts, be they expert falconers, novice, breeder or ornithologists.

Written following extensive research by the founders of the world's leading falconry portal, **Yarakweb.com,** Beatriz E. Candil García and Arjen E. Hartman, this work is a follow up to the award-winning book *El Gran Desconocido: The Red-tailed Hawk,* by Spanish falconer **Beatriz Candil García**.

ISBN 978-0-9555607-0-5

Book review by Peter Eldrett
The Falconer magazine (UK)

ARS ACCIPITRARIA
**by Beatriz E. Candil García
and Arjen E. Hartman**

Published by Yarak Publishing
ISBN 978-0-9555607-0-5

The front cover of this publication tells us it is 'An Essential Multilingual Dictionary for the Practice of Falconry and Hawking' and it does exactly what it says. A publication of over 280 pages, it is a full and comprehensive study concerning phrases and terms used in falconry in six different languages – English, Spanish, Italian, Dutch, German and French.

The dictionary, just like any other, is set out in alphabetical order with the main word or term to be described set in English and underneath a brief description of the meaning and then a table with that word or phrase in the five other languages.

After the acknowledgements and authors note, there is a short chapter called 'The Noble Art of Falconry: The Sport of Kings' which lays out a short history of the sport and a sub-section concerning falconry in literature with a quote from Shakespeare (Taming of the Shrew): "My falcon now is sharp and passing empty, And till she stoop she must not be full-gorged …"

Next comes the main part of the publication. The first and main section of the dictionary is titled 'Falconry terms and abbreviations' and contains the dictionary itself. It is laid out in simple form and is easy on the eye with just enough white space between entries without spacing out the entries just to fill the page.

As an example, Peregrine:- A widely distributed falcon popularly used for the practice of falconry which can achieve great speeds of more than 300 km/.

Then underneath is listed the five foreign descriptions of the main entry: Halcón peregrino (Spanish), Falco pellegrino (Italian), Slechtvalk (Dutch), Der Wanderfalke (German) and Faucon pèlerin (French).

There are so many other falconry related terms listed, from Aba to Yarak in this book, there are too many to list here and I find it hard to put the book down – trying to soak up the information contained on the pages.

The second section is titled 'Health and wellbeing' and is intended to be a quick reference for falconers not only on the different parts of a bird, but also the different diseases that a raptor can contract. It does not go into great technical detail but gives a good brief description of each ailment a bird can suffer: Avian Tuberculosis, Bumblefoot, Sour crop, etc. Viruses, Fungi, Parasites and other disorders are also listed.

The last two sections, Diurnal birds of prey and Nocturnal birds of prey list the scientific name of a species followed by the six equivalent translations of the name listed. There are even six indexes depending on which language you prefer.

I have to say that I find this dictionary a real joy to look at. It is simple in format and design and hats off to both authors who obviously have spent so much time compiling this edition. I recommend this book without any hesitation and I cannot think why nobody has thought of it before.

Here I am
And within the reach of my hands
She sounds asleep and she's sweeter now
Than the wildest dream could have seen her
And I Watch her slipping away

Though I know I'll be hunting high and low
High
There's no end to the lengths I'll go to
Hunting high and low
High

There's no end to lengths I'll go
To find her again
Upon this my dreams are depending
Through the dark
I sense the pounding of her heart
Next to mine
She's the sweetest love I could find
So I guess I'll be hunting high and low

High
There's no end to the lengths I'll go to
High and Low
High

Do you know what it means to love you…
I'm hunting high and low
And now she's telling me she's got to
go away
I'll always be hunting high and low
Hungry for you
Watch me tearing myself to pieces
Hunting high and low

High
There's no end to the lengths I'll go to
Oh, for you I'll be hunting high and
Low

Hunting High and Low, lyrics and music by A-Ha